Inigo Born, Rudolf Erich Raspe

**Travels Through the Bannat of Temeswar, Transylvania, and Hungary**

Inigo Born, Rudolf Erich Raspe

**Travels Through the Bannat of Temeswar, Transylvania, and Hungary**

ISBN/EAN: 9783743423909

Manufactured in Europe, USA, Canada, Australia, Japa

Cover: Foto ©Andreas Hilbeck / pixelio.de

Manufactured and distributed by brebook publishing software (www.brebook.com)

Inigo Born, Rudolf Erich Raspe

**Travels Through the Bannat of Temeswar, Transylvania, and Hungary**

# TRAVELS
THROUGH THE
# BANNAT
OF
# TEMESWAR,
# TRANSYLVANIA,
AND
# HUNGARY,

In the Year 1770.

DESCRIBED IN

A SERIES of LETTERS to PROF. FERBER,

ON THE

MINES and MOUNTAINS

Of these different Countries,

By BARON INIGO BORN,

Counsellor of the ROYAL MINES, in Bohemia.

To which is added,

JOHN JAMES FERBER's
MINERALOGICAL HISTORY of BOHEMIA.

---

TRANSLATED from the GERMAN,

With some explanatory Notes, and a Preface on the Mechanical Arts, the Art of Mining, and its present State and future Improvement,

By R. E. RASPE.

---

*In nova fert animus mutatas dicere formas.*

---

LONDON:

PRINTED BY J. MILLER, No. 6, OLD BAILEY;
FOR G. KEARSLEY, No. 46, FLEET-STREET.

MDCCLXXVII.

# PREFACE.

*HAVING* introduced Mr. Ferber's accounts of Italy, with some general views of those parts of Mineralogy, which have of late been improved, and may be further by a nearer examination of the Volcanoes and their various productions; it is but just that these accounts of the Hungarian and Bohemian mines should be accompanied with similar views on the art of mining; the nature of the different metallic mountains and of their various veins and productions.

This science has been the science of riches ever since the use of metals and of other fossils have been discovered, and turned to account by mankind. Though that discovery be a very old one, and the art of mining and of smelting be handed down to us by a long series of ages, and by different nations, some *of its scientifical parts are, however, brought to a less certainty than we might have expected.*

*Was it, that*

Mammon, the least erected spirit, that fell
From Heaven,

*was less adored, and his worship less followed than that of the fairer Muses? Was it*

That riches grow in Hell, that soil which best
Deserves the precious bane?

*Indeed it was not; for gold and silver, and riches have been in every age, in every clime, adored and pursued by all the nations, which had any claim to ingenuity, with such a zealous eagerness as would have done credit to any divinity. It is the common fate of the most useful and practical arts, to have been, in every age and in every nation, left in a state of infancy, and in the hands of working people, or of imposing quacks. Sovereigns have encouraged, and the wise and learned, with presumptuous attempts, pursued hazardous flights into the lofty regions of scholastic divinity and metaphysicks, beyond the reach of human abilities, and aimed at such objects, which in this world do not make us either wiser or happier, or richer or better. It is a very singular phænomenon in the history of mankind, that the arts of fortune-telling, of rhiming, of singing, fidling, reasoning, and speaking, should have been reduced*

*into*

*scientifical forms*; nay, that they should have been so highly improved, before any friend of men and good-sense thought of reducing the better arts of husbandry, of physic, of navigation, and of mining into the forms of sciences; of fixing them for ever, and of establishing them upon the evident and constant principles of nature. But such is the perverseness of human nature! Wants and accidents have co-operated to invent and to introduce the useful arts by the skill and ingenuity of some men, whom the Savages in the infant state of society have justly revered as their greatest benefactors, and as such have ranked and forgot them in the croud of their heroes and divinities. Ceres and Triptolemus, Pomona, Minerva and Esculapius, for being the inventors, improvers, or introducers of husbandry, gardening, weaving, and physic, have been by the savage Greeks, Tuscans, or Latins, consecrated to posterity by the same spirit of gratitude and veneration, which has sanctified the Evangelists, the Apostles, and the Saints amongst the Christians: but Poets, time, and human incongruity, what have they made in after-times of their glory, and of the arts and sciences which they taught? Let any one judge, who knows something of the history of mankind, whether the well-deserved reputation of their names has ever shone in its purest lustre, and whether their popular and salutary arts and sciences have ever been practised in that public-

*spirited*

*spirited benevolent manner, in which they had left or delivered them to mankind. Their names and their history have been involved in clouds of darkness and legends, and their arts and sciences by their will, the inheritance of all, have been engrossed by the selfish few. This happened in a very natural manner. Whether it happened necessarily, I will not determine; observing only, that making an exclusive trade of sciences and of arts, has never answered, and never will answer, the great and universal interest of mankind.*

*The most useful arts are precisely those which stand in an immediate connexion with the most general, most natural, and most indispensable wants of mankind. Their object is food, dress, and self-preservation. They must of course have been invented or practised by every family or society of men; and being on this account coeval with the first origin of mankind, their invention falls into the remotest antiquity of the primitive world, when a few natural wants of a few individuals, or of a few scattered families, could be satisfied by common ingenuity, in making use of the most obvious gifts and effects of nature, such as every climate afforded. In that primitive state, we see the arts of the Pecherais in Terra del Fuego, and of many little wandering tribes of men in almost every part of the world; and even the arts of civilized nations would be lowered again,*

*and*

*and turned down nearly to the same state, if by some sudden revolution they should happen to be at once deprived of the advantages of their climates, and exposed to the hardships and wants of other climates. Commodore Byron left and preserved on the coast of Chili, and the Russian sailors for many years left, and by their ingenuity and perseverance preserved, on the coast of Spitzbergen, will make good the assertion; and prove moreover, that wants and climate, going hand in hand, are the natural and first teachers of men, who for their vigour, ingenuity, and perfectibility, must be allowed to be lords of the world.*

*The self-invented arts of different nations, (and and why should not a similarity of wants or causes have produced a similarity of arts and remedies?) must for these reasons have been very simple, rude, and local in the beginning; that is, they must, under different climates, have appeared under different modifications. The canoes of some Indians, made of hollowed trees, speak a climate which produces plenty of timber; the canoes of the Greenlanders, made of seal-skins, and those of the Easter-Islanders in the South-Sea, being poorly made up, and sown together, of little bits of wood, speak a dismal want of wood. The form of the Chinese buildings and columns, is plainly that of the original tent of savages, wandering in warm climates, which produced light and slender*

*slender bamboo-trees.* The form of the Egyptian wonders of architecture seem plainly to tell, that the first inhabitants of that scorched climate, had cool rock-caverns to resort to for shelter; and that, when their increasing numbers attempted to imitate nature by art, they had little or no wood, but plenty of large rock masses. The old Indoes. being nearly under the same climate, seem to have built their most ancient Pagodas upon the same principles. The Greeks and Romans allow, that the Prototype of their most magnificent marble palaces was the original hut, made of timber, and even yet used in the milder climates of Asia. The dress of the Turks, Persians, Poles, and Hungarians, is of a different cut, but trimmed with fur, because they are offsprings of different nations in the northern parts of Asia, where dressing in fur is the advice and claim of the climate. The same original locality or nationality may be traced in the manipulations and technical words of the various arts of husbandry, hunting, fishing, fighting, and curing diseases; and as by so doing the origin of nations may be still more ascertained, and the invention of some arts pursued to their first beginning, it will likewise help us to feel, that many foreign arts have been to our cost introduced amongst us, in spite of the climate; and what here I am chiefly to insist upon, that the arts in the beginning must have been very simple and very rude.

*Wants*

*Wants in that infant state of society of hunters and fishermen, were presently and best removed by the simplest application of those natural effects or productions, which men experienced and saw before them. The causes of things, their investigation, methodising their accidental inventions, and fixing them for aftertimes, were then absolutely out of the question; and so they were even when the encreasing numbers and wants of these unsettled wanderers made the improvement of their original arts, or the introduction of new ones from abroad, the most acceptable gifts, which friends of mankind could bestow upon them. The Greeks were indeed but a very raw and ignorant people, when their gods and heroes may be supposed to have instructed them. So were the Britons and Germans, when their conquerors, the Romans, and the first Christian missionaries or apostles, acquainted them with the arts of making their life more comfortable. Each family held them as a treasure, and handed them to their descendants in a mere traditional manner. It is no wonder, therefore, that old tradition and history speak of Princesses skilful in the arts of the loom, of Sovereigns dressing their dinners, guiding the plow, and tending their herds. Her Highness Princess Nausicaa went to the river washing and scowering her linen; nay, even Reverend Abbots and Holy Priests are celebrated in the first ages of Christianity amongst* the Northern European nations, for having been skilful and laborious plowmen, gardeners, vintners, husbandmen, carpenters, joiners, painters, and physicians.

This traditional science of the arts was a natural consequence of the scattered, pastoral, and rural life, and it was attended with circumstances which proved no advantage to them. Being confined to single families, and their wants alone, their practice would but accidentally improve them, and these improvements were liable to be forgotten. Moreover, their drudgery must of course be left to the slaves, who for many ages, even in the politer nations, were employed to carry on the manual arts. Deprived of liberty and property, they were the more inclined to drudge on in a dull, stubborn, habitual manner, without any mind for improvement.

That nevertheless, the manual and mechanical arts amongst the Phenicians, Egyptians, Greeks, Carthaginians, and Romans, have been brought to a remarkable degree of perfection, was owing, not indeed to their slaves, but to the superior good sense and activity of their masters; to circumstances which produced a nearer connection of mankind in general; to wide extended navigation, commerce, and conquests; and finally, to a mercantile spirit and a culture of science, which have ever been the results and distinguishing blessings of human society, or

govern-

*government* brought to the highest degree of perfection.

Some of these reasons have at last rescued the manual and mechanical arts in Europe from the hands of bungling slaves, and brought them into the hands of free people; but that happy revolution has in most parts of Europe served the arts only by halves. It has been a great advantage to them; but having made more or less exclusive trades of them, they have been, and some of them are still kept as jobs and secrets, by short-sighted and narrow-minded mercantile selfishness.

This plainly appears by the ill-digested statutes and customs of many professions and trades, which, if possible, would be independent patent companies, at the expence of the whole; the very names of the art and mystery *of apothecaries, of clothworkers, of barbers, of cordwainers, and other trades, as expressed in the charters of their corporations at London, are striking instances to what lengths that unpatriotick selfishness has been carried in former times;* and even, if the word mystery *in these charters should be considered only as an equivocal, othographical blunder, instead of* mestier, *or* métier, *there are thousands of proofs that this old spirit of selfishness is yet alive, ever willing to take advantage of the knowledge of others, and never willing to promote it. Let us add the absurd contempt in which the proud*

*proud Barons and the self-conceited scholars have held formerly, and even yet hold, the greater part of the pretended servile and low mechanical arts, and we cannot wonder, that the progress and improvement has been so slow, and that many of them are still in a state of infancy.*

*It is only in the wisest and most enlightened ages, that we find some philosophers and wise men, stepping down from the giddy heights of their exalted station of learning, into which the barbarous ignorance of the vulgar and their own conceit had placed them, in order to fix, to rectify, and to improve the arts. Such ages produced amongst the Greeks and Romans, what Euclides, Hippocrates, Galen, Vitruvius, Columella, Cato, Pliny, Theophrastus, and some others, have left us on the arts; and it is in the true spirit of those glorious times, that after so many lost ages of scholastical dullness and mercantile selfishness, the Royal Academy at Paris, Mr. Chambers, Dr. Lewis, the Authors of the French Encyclopedy, and many friends of mankind in several parts of Europe, have undertaken of late to fix the various arts of mankind for after-times, and to establish them upon the principles of nature and mathematicks, better known at present than they ever were before. But various is their present state in different parts of Europe.*

*The* Art of War *is in these last two hundred years reduced in France, and especially in Germany,*
*upon*

upon so evident and scientifical theories, ascertained by practice, that these powerful empires must be the most happy of all, if tremendous armies, now and then methodically butchered, and the ambition of Sovereigns, flattered by conquest, did ensure them the blessings of peace, or any other blessing at all. There has been in those countries too much occasion for the improvement of this necessary and terrible art.

The Nautical Art in all its branches, on the contrary, is brought in England to the highest degree of perfection, because it is the kingdom of the seas; so are husbandry and numbers of mechanical arts and manufactories, because it enjoys the advantages of a plentiful soil, and of freedom in a higher degree than any other. Sed

Tu regere Imperio populos Britanne memento,
(Hae Tibi ERUNT artes) pacique imponere morem,
Parcere subjectis & debellare superbos.

But the Art of Mining, and its many subordinate branches, are in Germany, and its dependent countries, for various reasons, so highly improved, that for these last ages Germany has been justly considered as the most ancient and best school for miners. Though Tacitus, in his romantic account of Germany, told the Romans, that the Gods, either by a providential care, or by their dislike of the nation, seemed to have left the Germans unprovided with mines and metals,

*or*

or rather to have kept them till then unacquainted with their use and science; things have, however, since wondrously changed, both in respect to the mines, and in respect of their science. The greatest and richest chains and tracts of metallic mountains, which justly may be ranked with those in *Peru* and in *Hungary*, have been discovered there in a very remote antiquity, when the other kingdoms of *Europe* had scarce any idea of that kind of inland riches; and there have been ever since, and there are more mines and mountains yet actually working in *Germany* alone, than perhaps in all the other parts of *Europe* put together. Some mines on the *Rhine* and *Danube*, in *Lorrain*, *Alsace*, *Brisgow*, *Suevia*, and the ancient *Noricum*, seem to have been worked already in the decline of the ancient *Roman* empire. Many in the interior parts are reported to have been opened under the race of *Charlemain*. The mines in the *Rammelsberg* near *Gofslar*, and some of the adjacent ones in the *Harz-mountains*, belonging to the Electorate of *Hanover* and the Duchy of *Brunswick*, are fairly proved to have been discovered and worked to advantage as early as the middle of the tenth century (between A. C. 950. and 1000.) And the discovery of those in *Hassia*, *Misnia*, *Silesia*, *Moravia*, *Franconia*, *Tyrol*, *Steyermark*, *Carinthia*, and *Carniola*, cannot be supposed to have been much posterior in time to that of the former.

# PREFACE.

*To judge by the* technical language *of the German miners, washers, assayers, and melters, they do not seem to have learnt, or had their different arts from the Romans, or other foreign nations. It is downright German. It proves at least what I have shortly hinted before, that these arts are of very old standing in Germany; and as it is very compleat in every respect, and almost the same in the most distant provinces of Germany, it proves, that for a long series of ages these various arts have never been discontinued, and on that account they may be considered as national. Being by their very object and remarkable success naturally recommended to despotic Sovereigns, they have been very early favoured and taken notice of by the many legislators of Germany; and it must be owned, that the metallic general and particular laws of Germany, having been soon refined, have greatly contributed to keep these mining arts alive, by keeping the above mining countries in uninterrupted successful employment. And happy has it proved for Germany, as the inland parts of that extensive and populous country, without the working of these numerous mines, must have lost thousands of unemployed hands, and stand worse in the balance of trade than it hitherto is found to do. The mathematicks, mechanicks, hydraulicks, and the principles of chemistry, have been pretty early applied in Germany, to the traditional and empirical art of mining, as*

every

*every one may judge by the valuable writings of* Georg. Agricola, *(born* 1494——1555*) that excellent author of immense and practical erudition, who for these last* 250 *years has stood unparalleled and foremost amongst the classical authors on mining*; *and as during these last* 300 *years, Germany has produced a* Copernicus, Purbach, Kepler, Sturmius, Leibnitz, Wolf, Kaeftner, Meyer, Segner, Euler, Lambert; Albertus Magnus, *(born* 1193——1280*)* Paracelfus Theophraftus, *(born* 1493——1541*)* Sennert, Beccher, Kunckel, Stahl, Glauber, Hofmann, Juncker, Vogel, Marggraf, Model, Newmann, Cartheufer, Ercker, Cramer, Schlütter, Gellert, Lehmann, Poerner, Pott, Gerhard, Jufti, Waiz, Spielmann, *and* Meyer,——*besides many other* unmonumented *but great names in the history of mathematicks and of chemistry*; *it is not without some justice that foreign nations have considered the Germans as their masters in the art of mining, and not without some good reason, that the Germans have first endeavoured in their writings and academies to give this art that scientifical form which it is capable of, and to keep pace in it with the English, Swedes, Italians, and French, who of late have begun successfully to emulate their example.*

 *The object of this art is,*

 1. *The working and building of the mines in which they are found;*

2. *Their*

2. *Their extraction and separation from the ores and substances in which they are involved and mineralized;* and

3. *The investigation of fossil and metallic substances or ores.*

*Accordingly it is established upon different sciences, and may be divided into different parts.*

I. The Art of working and building the Mines *consists of a skilful application of natural philosophy and the mathematicks to this particular object; it is therefore to be divided into the following subordinate parts:*

  a. *The art of surveying and drawing mines.*

  b. *The art of breaking and blasting the rocks and veins.*

  c. *The art of timbering and building the works under ground.*

  d. *The art of correcting the air, which under ground, for many reasons, is liable to be damp, and unfit for respiration.*

  e. *The art of hydraulicks.* And,

  f. *At last, the art of mechanicks, for draining the mines of the subterraneous water, and for clearing them from the rubbish or ore by the various forces of nature, or various engines.*

*The old Egyptians, Greeks, and Romans, must be allowed to have not been deficient in these mathematical parts of the art of mining. Many of their subter-*

raneous buildings yet extant, and many of their great works of architecture, which ever will be objects of intelligent admiration, prove it beyond exception, and give credit to the ingenuity of their engines, which certainly we know but very imperfectly by their own written accounts. It would be extremely unfair to suppose that they had no engines; and that, unconcerned about the wretchedness of sentenced slaves, whom they employed in the mines as many Europeans employ the unsentenced innocent blacks, they left them unassisted by their ingenuity to every danger and hardship, which of course must befall the workmen, if they are led in the dark without intelligent guides, and condemned to do the various hard business of the mines by the strength of their hands, which we scarce are able to perform with the animated powers of nature enslaved by art. For unprejudiced mathematicians, or for intelligent antiquararians, it would be no hard task even yet to determine what degree of perfection they had actually attained in the above mathematical arts of mining. The Cloaca Maxima at Rome; the Emissario of the Lago Albano at Castel Gandolfo; their various aqueducts and cisterns; their colossal granite Obelisks cut in Upper Egypt, and thence transported as far as Rome; some mines in Transylvania supposed to be Roman works, and the watering engines, ever since the most distant antiquity, used in Egypt, would by

*an*

*an analytical examination, do justice to their ingenuity and unconquered spirit. But however astonishing they may appear to us in their works, and in the accounts and writings of Archimedes and Euclid, who would seriously pretend that the modern invention of the magnetical needle has not made the art of surveying under ground actually more certain and more easy than it was before? Who can deny, that the modern invention of gunpowder, and the art of blasting, has made us their masters? These powers of nature, which respectively lead us under ground, and arm us with the earthshaking strenth of Pluto and Neptune, were absolutely unknown to them. So were perhaps our various drawing and pump-mills, and our ventilators; so was the fire-engine, which is one of the most glorious monuments of English ingenuity, as, independent of the known powers of nature, it goes by a very active principle, in all the former ages scarce so much as noticed.*

II. *The art of* extracting *and* separating the Metals, *from the various heterogeneous substances, in which they are contained and mineralized, is carried on by water and fire, or by washers and smelters. It is therefore to be divided into the following subordinate arts:*

a. *The art of pounding the ores in mills.*
b. *The art of washing them.*

b                                          c. *The*

c. *The art of* metallurgy, *which, by the agents of fire and acids, separates, purifies, and respectively produces and destroys those various metallic and mineral substances, which are contained in the ores and fossil bodies, and are subservient and necessary to so many wants of human society.*

d. *The* Art of assaying *or* docimasy, *is rather a part of metallurgy, teaching, by small and nice assays of acids, fire, and weight, to determine the value, mixture, contents, and nature of the raw ores, or of the metals and mineral substances, produced by the greater operations of metallurgical furnaces and manufactories.*

If the Ancients, especially the Egyptians, Greeks, and Romans, must be allowed to have been pretty good empyrical metallurgists and smelters, as plainly appears by many of their works and accounts, it must, however, be allowed on the other side, that they have not left us any other, but perhaps a few traditional practices and processes, which, in respect to metallurgy and chemistry, are more vague and less to be depended upon, than in other more determined and evident arts and sciences. We can make but very little of their Hermes *and* Theophrastus, *and they had no* Archimedes *or* Euclid *for chemistry*; nor did they make use of it either in the preparation of their medicines, or in the examination of the elementary substances of nature. Their medicines were mixtures or decoctions

*decoctions of gross simples and substances, such as the vegetable, animal, and mineral kingdoms offered them; and their natural philosophy was, in respect to the elementary parts, but an ingenious guessing and reasoning in the dark, as it ever has been, and must be, without the assistance of chemistry, which resolves nature into its elements, and by acquainting us with many of their properties, unobservable and unobserved in their former combination, teaches how to make use of them, either in their concentrated simplified state, or in their new modelled combinations. It is to the Arabians that we are indebted for the advantages which philosophers, physicians, œconomists, and tradesmen, have reaped, and may reap, from this science, or rather from that scientifical chemistry, which we are at present possessed of, and have so much improved. Many of its technical names and its usual characters would prove it, if the writings of* Geber, Rhazes, *and many others, had left us any doubt. We must not, however, deprive our ancestors in Germany or England of their claim to the invention or use of more ancient metallurgical processes. I have mentioned already, that the discovery and working of the mines at Gofslar falls between the years* 950 *and* 1000, *after the age of* Geber *and* Rhazes, *who lived in the seventh and tenth century, but before the introduction of Arabian learning in Europe, which coincides with the crusades, and, to our knowledge, has produced no European chemists but in the beginning of the thirteenth century, when* Albertus Magnus, *or* Albert von Bolistaedt, *(born* 1193—1280*) and* Roger Bacon *(born* 1214—1294*) appeared. The metallurgical operations at* Gofslar *seem, therefore, in those earlier times, to have been established upon traditional processes, which were either Roman or German; and as, on account of the mixed irony and zincous refractory ores, these operations, though ever so much improved at present, are extremely various, compound, hard, and tedious, there is good reason to suppose, that even the traditional and empyrical science of the ancient German metallurgists was by no means inconsiderable. We have scarce any credible account that this traditional art should have been properly fixed for posterity, established upon scientifical principles, and remarkably improved by chemistry, earlier than the times of* George Agricola, *(*1555*) who for his valuable books* De Natura Fossilium *and* De Re Metallica, *deserves to be called the father of those many excellent chemical metallurgists whom Germany and other countries have produced ever since. I will not enlarge upon the dates and respective merits of* Agricola, Encelius, Erker, Becker, Stahl, Schlütter, Cramer, Gellert, Lehman, Vogel, Justi, Henckel, Pott, Marggraf, *and others; nor upon their many excellent disciples in* Sweden, France,

*France, and England, such as* Bacon, Rob. Boyle, Barba, Hellot, Macquer, Blake, Lewis, Woulfe, Beaumé, Sage, *and others; but I beg leave to observe, that metallurgy, being, upon the whole, and for the practical uses of the smelters, reduced upon pretty evident principles, is however very far from having attained its highest degree of perfection in respect to philosophical chemistry. Many mineral substances, ores, and fossils, are still very problematical; but the general spirit of enquiry spread over Europe bids fair to improve it in a quite different ratio from that in which it proceeded formerly. Mr.* Cramer's *new metallurgy, and Mr.* Delius's *proposals for copper-refining in Hungary, inserted in this publication, prove, that many metallurgical operations are capable of improvement by the principles of chemistry, duly applied; and the very principles of chemistry are at the eve of being better ascertained, and of being considered in a new light. At least very promising prospects have opened within these few years from Dr.* Priestley's *late experiments on air, and from the ingenuity and sagacity of Mess.* Pott *and* Marggraf *at Berlin, and of Mr.* Beaumé *and Mr.* Sage *at* Paris. *The former have seemingly acquainted us with new qualities and new sorts of air; but, properly speaking, they have exhibited to us only the phænomena of a new, active, subtile, elastic, and powerful solvent or menstruum hitherto not at all, or but imperfectly, made use of in our chemical analyses and assays. However, they must of course continue to enrich chemistry, and its dependent arts, sciences, and trades, with many valuable discoveries. I am confident the same must be the case with Mr.* Sage's *late and very ingenious theory and experiments on mineralization, which being above the understanding, or against the traditional creed of many chemical Virtuosi, have opened in France a new and ample field of abuse, and every where else an ample field of speculation, and of discoveries. It is with the new principles and discoveries in natural philosophy exactly as with the nostrums in physic. At first they are good* for every thing; *soon after old Method cries them down as good* for nothing; *but experience proves them at last to be good* for something.

III. *The* Art of investigating, discovering, and pursuing the metallic and mineral substances under ground, *is upon such terms as allow some hopes of establishing it upon certain principles.*

*Baron* Pabst v. Ohain's *idea of a subterranean geography, seems to imply that he thought of it; and the mineralogical accounts of Mess.* Ferber, *Baron* Born, *and others, prove to me, that such an art* may be *invented, and likewise that its invention is in some forwardness,*

I shall not speak of Chance, *that great discoverer of mines, formerly worked and yet working;* nor shall

shall I enlarge upon the virgula divinatoria, or divining rod, *tried by philosophers in England, even so late as the times of* Robert Boyle, *and not an hundred years ago seriously applied in France for the discovering of mines, treasures, wells, robbers, and murtherers. We do not know how to methodize the former, and we are fully convinced that the latter has never answered any purpose but that of making dupes. They are, therefore, best left and recommended to ignorant people, who delight in* darkness visible.

*The only principles, upon which this very interesting art may and must be established, are* Mineralogy *and* Oryctology.

Mineralogy, *or a sufficient historical knowledge of the fossil bodies, is of absolute necessity to the miner and to the learned. It acquaints the former with the name, form, colour, texture, appearance, value, and other properties of the fossils, and it makes science intelligible by scientifical, determined names of their characteristic properties. For want of this science, quantities of rich ores and fossil substances have been formerly thrown amongst the rubbish of the bingsteads; and there is scarce a mining country in which they have not some time or other paved their highways with stones and rocks of value. I know, from very respectable authority, that that was formerly the case of the Cobalt-ores in Hesse, which at*

*present*

*present produce an annual revenue of about* 14,000 *l. clear of expences. That the deficiency of languages in that part of the art of mining, which treats of fossils, has been hitherto a great obstruction to its improvement, will not be denied, and has been severely felt by every one who wishes to instruct, or to be instructed, in mining, metallurgy, chemistry, and natural history. Only a very few fossil substances have determined names in common life, and in the languages of the politest nations. Such are the purer and finer metals, some salts, and some stones. The infinite variety of their mixtures, different state, mineralization, chemical properties and affinities of their colour, form, hardness, weight, situation, native place and origin, if understood or noticed by the miners and metallurgists, are expressed either by technical or by provincial names, which, to the generality of men, or to foreigners, are what formerly the Greek was to the Monks. I beg leave to observe, that they have not been hitherto taken sufficient notice of in those numerous mineralogical systems, which have appeared these last fifty years. Their authors consider mineralogy under too confined points of view; and many of them have indulged themselves in new and very often arbitrary names and idle classifications; so that an egregious and nearly Babylonian confusion has been added to the old deficiency of languages; and that science, upon the whole, has been less benefitted by them*

*them than justly might have been expected.* It must be acknowledged, however, that the chemical classification, and the nomenclature of fossils, introduced into mineralogy ever since the first appearance of Pott's Lithogenesy, *and the systems of* Wallerius *and* Cronstedt, *have been great advantages to science. Being established upon their constituent parts, and upon reality, they may, under certain allowances, prevent further obscurity and confusion, and perfectly answer the views of chemists and metallurgists. But as they stand at present, can they fully answer the expectation of miners, of natural philosophers, and of friends to science? They must be the basis of mineralogy; no competent judge, and no man of sense, will dispute it; but to benefit the miner, they should be explained in his own technical or provincial language, which is generally neglected; and to satisfy the natural philosopher, they should be established only upon evident principles of chemistry, and never presume to classify fossil substances, which are not hitherto sufficiently examined. Moreover, many fossils are so compound in their mixtures, that a chemist may make any thing of them; and most part of their chemical characters are so far abstruse and obscure, as they relate rather to a future state than to that raw and natural one, in which we see and discover them. A chemical mineralogist will at most tell you what fossils are good for, or what you may make of them,*

them, *or of what they are composed. He will thence make you guess likewise by what natural operations they may have received that raw and natural form in which we find them. But fire, crucibles, retorts, alembics, acids, and touchstones, are insufficient to teach with certainty by what natural operations they really were produced. These are facts, which chemical principles and good reasoning will and must explain; their reality must be ascertained by historical evidence, or by ocular inspection and experiment, for the same mixture of fossils can be produced by fire and by water, by melting and by solution, by sublimation, by precipitation, and by other operations. This we plainly experience in our laboratories; and as these operations are really different from each other, and generally productive of particular forms and circumstances; it is but just to suppose, that the same or similar, operations in the great laboratories of Nature, are and must be productive of similar forms and circumstances, and that the particular forms and circumstances in which the fossils are found, should be nicely noticed by mineralogists, who pretend to give full information and adequate ideas. They must henceforth examine, rank, and describe them not only as individual substances by their chemical properties, colour, texture, and form, but they must consider them likewise under the more extensive point of view of their former natural situation, stratica-tion,*

tion, connection, and vicinity with other fossils in the native places, beds, and veins in which they are found. By so doing, mineralogy receives a latitude, of which it hitherto has been deprived; fossils appear in their only and true natural order, which is that of their chemical properties and of their natural situation; the fossil beds and veins become monuments of former revolutions, or of natural chemical operations; and, in short, the art of discovering or pursuing them under ground will be reduced into probable rules. I do not indulge a chimerical fondness for a favourite idea. I have traced the outline of a mineralogical system upon these principles; that is to say, a system which goes hand in hand with the principles of chemistry and of stratification; and I must be false to truth, if, for an ill-placed bashfulness and modesty, I should not publicly acknowledge and recommend its advantages. Hints of this new mineralogy are thrown out in my account of the German volcanoes, and especially in my preface and index to Ferber's Letters on Italy, and to this publication. I hope they will suffice for the intelligent, and make good the assertion, " that the invention of the art of discovering mines is in good forwardness."

Philosophers, ancient and modern, have hitherto considered mountains in general from a point of view which was too confined, or entirely different from that

that of mineralogy and mining. For being unimproved by the light of volcanoes, and by that extensive knowledge which they might have reaped in the deepest mines of the highest mountains, and from the instruction of unscientific miners, they stuck only to their libraries, and to the uppermost crust of the earth, which, without any great trouble to themselves, they had an opportunity of examining in the most pleasing countries, and in the most superficial quarries of sand-stone, limestone, and slate. We are not to wonder, therefore, that orology, or the science of mountains, is so little understood amongst the learned; and that the descriptions of the higher mountains in Peru, Teneriffa, Switzerland, and different parts of Europe are generally filled with meteorological observations, botany, and other accounts, which leave their very nature in a minerological and orological respect full as unknown as they were before. The consequence was plain, that general conclusions have been too rashly drawn from a single kind of mountains, and that the pretended systems of the origin of the mountains in general are, for the greater part, so very romantic and superficial.

Experience and history prove the mountains and strata of the earth to be of a very different nature, origin, and antiquity. They have been accordingly divided by Mr. Giovanni Arduino, at Venice, into primitive, secundary, modern, (Tertiarii) and volcanic mountains. I shall not repeat from Mr. Ferber's Letters what

*what characters he has given of these various mountains; but I must notice, that* primitive mountains, or strata, *have been spoken of very early by other philosophers, and that they are generally understood to consist of simple rocks, less stratified than the incumbent strata, and never containing in their paste and mixture any petrefactions, of adventitious, organic bodies of shells, other animals or plants. That there are such mountains and strata, is unquestionable, because they are found either at the bottom of the deepest mines, or bare appearing through a variety of incumbent other strata at the summits of the higher mountains. But it would be extremely presumptuous to insist upon their being true primitive mountains. The only consequence which fairly can be drawn from Mr.* Ferber's *and Mr.* Arduini's *Observations in his Raccolta di Memorie Chimico-mineralogiche : Venezia* 1775) *is, that the reddish granite (granito rosso d'Egitto) and the micaceous and horn-slate, for being every where found below a variety of other incumbent strata, must, in respect to time and origin, be anterior and different from them, and that, for this reason, they may justly be called the most antient rocks hitherto known.*

*The* secundary strata and mountains, *chiefly consisting of limestone and argillaceous slate, are accumulated on the former, and the* more modern ones *are incumbent on these. They owe their origin to a variety*

*variety of causes and accidents, as may be proved even by history.* Nor are even the volcanic mountains and strata *produced by and under the same circumstances, as sufficiently is hinted by me in the preface to Ferber's Letters, and in my account of the German volcanoes.* Of course they must all of them, for chemical and historical reasons, offer a variety of circumstances in the form, mixture, and substance of their paste and rocks, in the situation of their strata, in the nature and direction of their veins, and in the nature and mixture of the parasitical rocks, which are produced in their veins and caverns.

The miners in Germany, whose ideas have been generally confined to their main object, and to the nature of the mountains, in which they worked, have, instead of the above division of the mountains, divided them into flat *and into* gang *mountains.* (Flots, and gang-geburge.)

*By the former they understand stratified modern mountains, which generally surround the higher and more ancient ones, and are worked not for their veins, but for the contents of their strata, which are less dipping, and more horizontal, than those in the higher mountains. Such are the slate copper-works in Mansfield, the coal mines in general, and many iron mines.*

*By* gang-mountains, *they understand higher metalic mountains, which are working for their* veins or stocks, *and consist in Germany, Bohemia, and Hungary,*

*gary, of granite or micaceous and horn slate, or of what naturalists would call primitive mountains.* By gang, or gang-geburge, *they understand likewise those various substances which either do never appear in stratified rocks, but in veins (or gaengen) only, viz. the whole tribe of parasitical stones, of quartz, spar, fluor,* &c. *or those, which in particular cases are found to fill veins, joints, and stocks, as granite, slate, zinnopel, grit, clay, and other matrices of metals and minerals. In this sense, it is used to distinguish them from the rocks of the mountain on both sides of the vein or stock; and it is absolutely a relative denomination, since the same substance may be in some places particular to the vein, and in some others to the mountain.*

*Father* George Agricola *is undoubtedly the first, and, I dare say, till of very late, unparalleled in respect to some scientifical knowledge of the veins, their run, and their rules. What he knew and drew of it, he knew from the miners; but as ever since they have scarce been consulted at all, by philosophers who attempted to create and to dream mountains and worlds, and systems of mountains and worlds, it is no wonder that hitherto the learned should have so little added to that stock of science, which he has left us. The best general accounts, besides his, are* Lehman's *two German Treatises* on the Flat Mountains *and* the Metallic Matrices, *and some general principles in those valuable* elements of the art of mining, which

*which have been published by the academy for miners since* 1765, *established at Freiberg in Saxony. But we are very far from being thoroughly acquainted with their nature. We know them only by a few good observations made in Germany, Sweden, and Hungary. By these it appears*

I. *That the veins of the same mountain, nay, of very extensive tracts of land, are subject to the same rules in respect to their direction, dipping and crossjoints, and of the same nature in respect to their contents, ores and fossils.*

II. *That veins of the same mountain, running in the same direction, and through and under the same rocks, are loaded with the same ores and fossils; and accordingly seem to have been produced and loaded by the same natural revolution.*

III. *That those which cross them in contrary directions, seem very often to have been produced in different times, and by different revolutions, when they are loaded with different substances.*

IV. *That the veins of the incumbent mountains, for example, of calcareous or slate mountains, have their particular rule of direction, dipping, cross-joints, and contents, very often unaffected by the different rule of those veins, which are in the lower rocks of slate or granite; whence it appears, that these latter were, in point of time, anterior to the former.*

V. *That*

V. *That the veins commonly turn quicker and richest in the crossings.*

VI. *That ores and metals are produced in the crossings, which did not appear before either in the main vein, or in the cross-joints.*

VII. *That they are generally quick when running or dipping along, and between the limits of different adjacent or incumbent rocks; for example, in the limits of granite and incumbent slate, or in those of slate and incumbent limestone.*

VIII. *That their contents or loads of rocks, parasitical stones, ores, and metals, generally have a natural and chemical relation to the rocks in and under which they are running; that is to say, that veins in granite carry tin, wolfram, pyrites, black-lead, quartz, and granite grit; that veins in limestone carry spar and fluor, besides lead and other metals; that veins in the Hungarian metallic rock are filled with quartz, fieldspath, gold and silver; that veins in hornstone produce gold, silver, and zinnopel; that veins in slate are loaded with argillaceous substances, quartz, silver, lead, copper, and iron; and that the natural productions of limestone, slate, porphyry, trapp, and volcanic beds, if incumbent on deeper veins, appear in them, or produce modifications of fossils, which their own matrix, rocks, or sides, would*

c

not have produced by themselves. To verify these assertions, I refer the intelligent reader to a tabellary abstract of some mines described in this publication, and beg leave to observe, that, by similar abstracts of all the good mineralogical and orological accounts which are come to my hands, I am enabled to give in a new edition of my System of the Earth, something more satisfactory about orology and the metallic mines in general than hitherto has been given to the public. They are the work of many years, and of great labour; and with the various improvements of the above System, not undeserving that generous support and encouragement which, though a foreigner, I make bold to expect from unprejudiced friends of science, and from those gentlemen, whose interest and business it is to see somewhat clear under their own ground, and to prevent those many impositions and disappointments, to which adventurous, unprincipled miners are subject.

I. *This System of the Earth and Mountains is to appear in two volumes in 4to. and to contain*

II. *An exact description of the surface of the earth and its strata; with an appendix of my own, and various of the best orological observations reduced into tabellary forms.*

III. *His-*

III. *Authenticated accounts of the various revolutions which have produced, changed, and affected the mountains, strata, and veins.*

IV. *Candid and literary accounts of the best orological systems, especially of that of the Greeks, and of the late Robert Hooke, who did not live to give it its due extent.*

V. *A short explication of the phænomena on the surface of the earth, by the above historical accounts, supported by the principles of chemistry.*

VI. *An ample sketch of a new system of mineralogy for miners, laid down upon the principles of metallurgy and stratification, with a constant reference to the technical and provincial language of the miners and smelters.*

VII. *Some scientifical sections, plans, and maps, beside some instructive and ornamental drawings of unnoticed fossils and petrifactions, will be added; and as I must be at the expence of some enquiries in the mining countries of Great Britain and Ireland before I can put the last hand to the work, which will not be without some expence, I desire the friends of science to take this work under their protection, to leave their names and orders at Mr. George Kearsley's, bookseller, in Fleet-street; who likewise will take care of their commands and enquiries, directed to me, in whatever*

*whatever I may be useful to them. More particular proposals will be published as soon as the subscribed or ordered copies amount to a certain number. Meanwhile*

> Current utiliter mei
> Nullo cum strepitu dies;

*and I shall continue the publication of other valuable mineralogical and orological tracts, which, for the better convenience of the purchasers, will appear in the same form as this, and as Ferber's Letters. In the next I intend to lay before the English public*

1. *Supplements to the* Mineralogical Letters *of Baron Born, taken from an improved edition, which lately has appeared in Germany.*

2. *Abstracts from* Gio. Arduino's *Raccolta di Memorie Chimico-mineralogiche, Metallurgiche e Orittografiche, published in* 1775 *at Venice.*

3. Ferber's *Accounts of the Mines in Derbyshire, published in* 1776 *in Germany.*

4. Ferber's *Accounts of the Mines in the Palatinate.*

5. *Ab-*

5. *Abstracts from Mr.* Colini's *Mineralogical Travels in the Palatinate; reserving for other future publications the best mineralogical accounts of the mines in the Harzforest,* Saxonia, Hesse, Tyrol, Sweden, *and* Italy.

R. E. RASPE.

London, Sept. 1776.

P. S. *The ounce spoken of in this publication is equal to one half ounce English, and the annexed orological tables are given only as essay, not as compleat abstracts of the accounts contained in this volume. Moreover I beg to observe, that the smallness of the size would not allow to specify the different species of ore, nor to give some other minute details.* Sed sapienti sat!

# APPENDIX TO THE PREFACE,
## OROLOGICAL TABLES OF SOME MINES DESCRIBED IN FERBER's LETTERS FROM ITALY, AND IN THIS PUBLICATION.

Thofe of the MINES at JOACHIMSTHAL, in Bohemia, to be compared with the MAP.

| Name of the Place and Mine. | Mountain, or Sides of the Vein. | VEIN | | | |
|---|---|---|---|---|---|
| | | Rock. | Ore. | Direction. | Dipping. |
| Taiftriz, near Pegaw, in Steyermark. Ferber's Ital. | Limeftone blue argillaceous flate. | Spar and quartz. | Leadglance, with filver. | | |
| Idria, in Crain. Ferb. Ital. Lett. | Limeftone blue arg. flate. | Slate. | Cinnabar and quickfilver. | | |
| Feltrino. Ferb. | Limeftone flate. | | Quickfilver. | | |
| Schio, in Monte Trifa. Ferb. | Limeftone alternating with volcanic strata. | Spar. | Silver, lead, copper, manganefe. | | |
| Val'e di Gorno, in Bergamafco. Ferb. | Limeftone and lava. | | Lead and blende. | | |
| Montieri. Ferb. | Limeftone and flate. | Spar and quartz. | Silver, lead, copper, iron. | | |

| CROSS VEINS, or CROSS JOINTS. | | | | | | | |
|---|---|---|---|---|---|---|---|
| IMPROVING. | | | | CUTTING OF, OR STRIKING DEAF. | | | |
| Rock. | Ore. | Direction | Dipping. | Rock. | Ore. | Direction. | Dipping. |

| Name of the Place and Mine. | Mountain, or Sides of the Vein. | VEIN. | | | |
|---|---|---|---|---|---|
| | | Rock. | Ore. | Direction. | Dipping. |
| Nagyag. Born. | Red clay, sandstone, metallic rock. | Feldspath and fat quartz. | Gold and silver mineralized; auriferous antimony; arsenic; cinnabar. | South to North. All the quick veins parallel. | West to East. |
| Toplitza. Born. | Clay slate, metallic rock. | Quartz. | Gold, red silver ore, lead. | South to North. | |
| Fuezes. Born. | Clay slate, metallic rock. | Quartz. | Gold. | South to North. | |
| Felso Banya. Born. | | | | | |
| Borkul mine. | Hornstone. | Zinnopel. | Gold and silver. | | |
| Great Mine. | Hornstone and metallic rock. | Zinnopel. | Gold and silver. | | |
| Smolniz. Born. | Micaceous and blue slate. | Clay and quartz. | Copper pyrites, and silver. | East to West in hour 6. Three quick veins parallel. | In 75 degrees. |

## CROSS VEINS, or CROSS JOINTS.

| | IMPROVING. | | | | CUTTING OF, OR STRIKING DEAF. | | |
|---|---|---|---|---|---|---|---|
| Rock. | Ore. | Direction | Dipping | Rock. | Ore. | Direction. | Dipping. |
| Quartz & fluor. | Sulphur, antimony, manganese. | In the hanging. | | | | In the hading, crossing the main vein in an acute angle. | |
| | | East to West. Bring the veins in the hading, and quicken them. | In 75°. | | | West to South in hour 9, or hour 21. Bring the veins in the hanging, and strike them deaf. | West to East, or to North. |

| Name of the Place and Mine. | Mountain, or Sides of the Vein. | VEIN. | | | |
|---|---|---|---|---|---|
| | | Rock. | Ore. | Direction. | Dipping. |
| *Skemniz.* Born. Spitaler Vein. | Clay slate, metallic rock. | Quartz and zinnopel. | Gold, silver, lead. | North to South, or South to North, between hour 12 and 4. | West to East, between 30′ and 70′. |
| S. John's. | Metallic rock. | White clay and quartz in the hanging; zinnopel in the hading | Silver. | In the hanging of the former; and parallel. | |
| Beaverstoln. | Metallic rock. | Quartz, zinnopel, spar | Gold, silver, lead. | North to South as above, between 12 and 4. | West to East, between 30′ and 70′. |
| Theresia. | Metallic rock. | Zinnopel. | Gold, silver, lead. | North to South. | East to West; then vertical; at last West to East. |
| *Catharinaberg,* in Bohemia. Ferber. | Gneiss. | Gneiss or grit of granite. | Silver, copper; mineralized and native. | South to North, or North to South, in hour 2. | Between 60′ and 90′. |
| *Pressniz.* Ferb. Maria Kirchbaw. | Gneiss. | Gypseous spar. | Silver. | North to South, in hour 12 and 1. | |

## CROSS VEINS, or CROSS JOINTS.

| IMPROVING | | | | CUTTING OF, OR STRIKING DEAF. | | | |
|---|---|---|---|---|---|---|---|
| Rock. | Ore. | Direction | Dipping | Rock. | Ore. | Direction. | Dipping. |
| White clay, spar, quartz. | Silver. | In the hanging | | | | | |
| White clay, quartz, spar. | | East to West. Quicken the veins. | | Coarse clay, spar | | | |

|  | Name of the Place and Mine. | Mountain, or Sides of the Vein. | Rock. | Ore. | VEIN. Direction. | Dipping. |
|---|---|---|---|---|---|---|
| **NORTHERN VEINS** | *Joachimsthal,* in Bohemia | | | | | |
| | 1, Gold-rose hading | Slate, grey micaceous | Clay; red hornstone or flint; clay slate; spar; quartz. | Silver, lead, cobalt, arsenic: rich ores. | South to North, h 1. 6¼ p | East to West, between 54° 78°. |
| | 2, Gold-rose hanging. | Ditto. | | | h. 12. 5½. | Ditto. |
| | 3, Fund-grub. | Ditto. | Ditto. | Ditto. | h. 12. 6½. | Ditto. |
| | 4, Baker's Vein. | Ditto. | Ditto. | Ditto. | h. 7. 7½. | Ditto. |
| | 5, Geshi-eber. | Ditto. | Ditto. | Ditto, silver native, in the cross | h. 10. 4 p | Ditto. |
| | 6, Rose from Jericho | Ditto. | Ditto, and red spar, in the cross. | Ditto, and glass and red silver ore: in the cross. | h. 2. 3 p. | Ditto. |
| | 7, Sweitzer. | Ditto. | Ditto. | Ditto. | h. 1. 2¼. | Ditto. |
| | 8, Young Sweitzer. | Ditto. | Ditto. | Ditto, and native silver and glass ore N. B. The ores, and the richer ones, chiefly in the crosses of the following eastern veins. | h. 2. 4¼. | |

## CROSS VEINS, or CROSS JOINTS.

| | IMPROVING. | | | | CUTTING OF, OR STRIKING DEAF. | | | |
|---|---|---|---|---|---|---|---|---|
| Rock. | Ore. | Direction | Dipping | | Rock. | Ore. | Direction | Dipping. |
| Trapp. | | East to West. | South to North | | | | | |
| Trapp. | | Ditto. | Ditto. | | | | | |
| Trapp. | | Ditto. | Ditto. | | | | | |
| Porphyry, fat clay | | South to North | | | | | | |
| All the above Northern veins are besides constantly improved by the crossings of the Eastern ones, which run from East to West. | | | | | Porphyry | | South to North | |
| | | | | | Porphyry | | South to North | |

| Names of the Place and Mine. | Mountain, or Sides of the Vein | VEIN | | | |
|---|---|---|---|---|---|
| | | Rock. | Ore. | Direction. | Dipping. |
| **Joachimsthal, in Bohemia** | | | | | |
| 1, Lawrence. | Slate gray micaceous | Clay, clay slate, spar, and quartz. | Silver, lead, cobalt, arsenic: rich ores. | East to West, h. 5. 1¼ p. | South to North, between 60. 73'. |
| 2, Susanna. | Ditto. | Ditto. | Ditto. | h. 6 ¼. | Ditto. |
| EASTERN VEINS. 3, Ursula | Ditto. | Ditto. | Ditto. | h. 6 6¼ p. | Ditto. |
| 4, Andreas. | Ditto. | Ditto. | Ditto. | h 7 2¼ p | Ditto. |
| 5, Cow Vein | Ditto. | Ditto. | Ditto, and silver native glat ore, lead glance. | h. 7. | Ditto. |
| 6, Rose Vein | Ditto. | Ditto. | Ditto. | h. 6 ¾ p | Ditto. |
| 7, Elias. | Ditto. | Ditto. | Ditto. | h 7 ½ p | Ditto. |
| 8, George stoln. | Ditto. | Ditto. | Ditto. N.B. The ores and the richer ones chiefly in the crosses of the above Northern veins. | h. 6 3¼ p | Ditto. |

| CROSS VEINS, or CROSS JOINTS. |||||||||
|---|---|---|---|---|---|---|---|
| IMPROVING. |||| CUTTING OF, OR STRIKING DEAF. ||||
| Rock. | Ore. | Direction | Dipping. | Rock. | Ore. | Direction | Dipping. |
| | | | | Trapp. Porphyry Trapp. | | East to W South to N. South to N. | South to N. East to W. |
| | | | | Trapp. Porphyry | | East to W South to N | South to N. |
| | | | | Porphyry | | South to N. | |

N. B. The above Northern veins cross and improve the Eastern ones.

# TRAVELS

THROUGH THE BANNAT OF

TEMESWAR, &c.

## LETTER I.

*Temeswar, June* 14, 1770.

MY journey from *Shemniz* to this place has scarce offered me any object that might make this letter agreeable to a naturalist of your cast. Had I, besides my little mineralogical science, some knowledge in Botany, my three days travelling over barren heaths from *Ofen* to *Segedin*, and thence to *Temeswar*, might have perhaps procured me an opportunity to entertain you at least with the names and descriptions of some plants. But alas! I am no Botanist, tho' that is not my fault. You well know how fond I am of natural history. But I never met with any proper opportunity

tunity to improve in this part of science. Except at *Vienna*, there is no academy in all the *Austrian* states, in which Botany is taught; nay, even at *Vienna*, there is no Professor of Natural History. For this reason you need not be astonished that natural history is entirely unnoticed and neglected in *Austria*, while the *English*, *French*, *Swedes* and *Russians*, for the sake of useful science, examine their own and the remotest countries of the world. But to what purpose these complaints? You may guess by them the dissatisfaction, which will attend me on my journey through the mountains of *Bannat*, *Transylvania*, and part of the *Carpathian hills*. All the riches of *Flora*, during the finest season of the year, displayed in these parts, will be scarce at all enjoyed by me. However, I do what I am able to do, and I repeat my former promises, that you shall have a share of the minerals which I collect, and accounts of the nature of the mountains, and the working of the mines, which perhaps may be new to you.

From *Shemniz* to *Ofen* the mountains consist of the same argillaceous rock, which is mixed with quarz, sherl and mica, and composes the whole mass of mountains about *Kremniz* and *Shemniz*. In some places, and especially at *Deutsch-Pilsen*, they have likewise discovered some copper and silver-veins, drained some old and drove some

new

new galleries; but to no great advantage. All these mountains are covered with argillaceous slate and limestone.

Near *Waizen*, a handsome little city on the *Danube*, begins the plain, which uninterruptedly stretches thence to *Temeswar*, and to the left hand to *Debreczin*, and the limits of *Transsylvania*. In three hours I came to

*Pest*, where I spent a day. This city, adorned with magnificent structures in the newest taste, is entirely built of petrifactions. The quarry, whence they fetch the stones, is near *Ofen*, a city directly opposite on the other side of the *Danube*. I examined these calcareous hills, productive of the best wine of *Ofen*. They consist of a porous limestone, which is filled with innumerable quantities of chamites, turbinites and pectinites. Our *Walch's*, *Schrotters* and *Hupsches*, with several other gentlemen of that kind, who are affraid of coal dust, and the horrors of smutty mines, and hunt after petrifactions only on the surface of the earth, might in this place make rich crops; nay, they might perhaps, from this immense stock of shells, pick up some chamites or pectinites, with some unknown undescribed stripes, wrinkles, folds, warts and points; and then, mercy upon our ears! how they would indulge themselves in God knows what analogy or similarity; in forming far fetched names,

names, and singing forth the praises of their important discoveries! To us simple mineralogical folks it is sufficient to have found here marks of an ancient sea's covering this part of *Europe.*

The hot baths at *Ofen* are spoken of by all geographers. Mr. *Laurentius Stocker* describes them at large in his *Thermographia Budensi.* According to his account, their constituent parts are sulphur, lime, and iron.

Beyond *Ofen* begins the famous *Ketskemite-heath.* It is all over covered with grit sand (glarea Linnaei) mixed with broken sea shells. The stones which now and then appear straggling, are ferruminated by this sand. I travelled often six hours and longer without meeting with any tree or house, except the stage houses. However this plain, fifty *German* miles square, feeds vast quantities of cattle. Near *Debreczin* they dig out of some swampy grounds of this heath the *Sal alcali minerale nativum,* mixed with some clay. For many years they have made of it the excellent *Debreczine* soap, which sells over the whole kingdom. In former times they considered this as a common saliter. Mr. *Stephen Wefzpremi,* a celebrated physician at *Debreczin,* and Mr. *Just John Torkos,* were the first who examined it. The former spoke of it in his

*Tentamine de inoculanda peste. Londini,* 1755.

and

and the latter in his treatise

*De sale minerali alcalino nativo Pannonico. Posonii*,
1763.

I heard lately from *Vienna*, that a young physician, Mr. *Gabriel Pazmandi*, from *Comorra* in *Hungary*, has published a new treatise on this salt, its native situation, qualities and powers.

I observed on this heath some flocks of large eagles, and some birds in the swamps, which were unknown to me, and may be perhaps for want of a proper description, or a scentifical zoologist to observe them, uninserted and unnoticed in the systematical catalogues of birds.

Beyond the *Theissa* (*Tibiscus*) and as soon as I left *Turkish Carisha*, the soil appeared richer and more entertaining. Here are plantations of trees, corn-fields, and plenty of colonies, whose establishment costs to our imperial queen immense sums annually.

The villages are built upon a regular plan; the houses, for want of wood, built of unbaked bricks, and thatch'd with reed (arundo.) They have generally a parson, a school, a corn magazine, and an accountant or inspector. Every colonist receives at his arrival a suitable house, the tools of husbandry, the household implements, some horses, and a piece of ground. After some years he gives the tithe of his crop as a contribution, and then

then he may pay every year what he can afford of the whole property.

A good huſbandman is ſure to proſper here. Perhaps it might have been made more eaſy to them if the villages had been planned ſmaller. There are ſome that contain 3 or 400 houſes. As every coloniſt is poſſeſſed of a large waſte ground, which he is to cultivate, many of them have an hour's ride before they can reach it.

## LETTER II.

*Temeswar, June* 17, 1770.

YOU know that two years ago I travelled in this country; besides I was born in *Transylvania*. I have therefore materials for a letter, which may for the want of natural history, if not please, at least entertain you.

The *Bannat* of *Temeswar* is that tract of land in *Hungary*, which in the *Homannian* maps is found under the title of the *Csanader* or *Temeser* county. It is under the 45th degree northern latitude, is 22 *German* miles in length, and 15 or 16 in breadth. Its boundaries are to the north the river *Maros*, to the west the *Theissa*, to the south the *Danube*, and to the east tremendous chains of rocks, which separate it from *Transsylvania* and the greater *Wallachia*. But on this side it joins to the continent; in respect of the other sides it is a peninsula. It is divided into eleven districts or bailiwicks, viz. that of

*Csanad*, of *Czakova*, of *Szent Andrash*, of *Szent Miklosh*, of *Beczkerek*, of *Uy Palanka*, of *Vershez*, of *Orsova*, of *Caransebez*, of *Lugosh*, and of *Lippova*. Every district is subdivided into smaller jurisdictions,

risdictions, which are called processes. A bailiwick consists of the bailiff, a comptroller, two or three under bailiffs, a scrivener, some advocates and upper-kneses, which are a sort of national magistrates. All these bailiwicks are immediately under the country-administration, and this under the royal court chamber-deputation at *Vienna*. The *Bannat* being a domanial estate of her majesty, is entirely independent of the *Hungarian* states. The chief town and the center of the country is *Temeswar*, a regular, fine, and strong place, but unwholesome on account of its swampy situation. Agues and inflammatory fevers of all kinds rage here every season, and procure to the physicians uninterrupted business.

Here is the general government, the country administration, the provincial court, the chapter of *Csanad*, whose bishop is by his own right *primus inter pares* in this country, and two patentee-commercial companies for the *Austrian* sea-ports in Italy. The whole eastern part of the country is mountainous and best inhabited; the western part is flat and swampy. In this are large uncultivated plains, which government takes care to plant with *German* colonies from the *Swabian* and *Rhinish* circles. On the four corners of the country are some strong places, such as *Canisha*, *Semlin*,

*Semlin*, *Mehadia*, and *Lippa*. *Szegedin* and *Arrad*, situated on the other side of the *Maros* and *Theiſſa*, are *Hungarian* dependencies. None of these four places are remarkably strong. However, they are celebrated in the history of the *Turkiſh* wars, as are likewise *Panſowa*, *Uy-Palanka*, and *Orſowa*. The rivers in the *Bannat* are of no importance, as running only through a short tract of land; but the *Temes* and *Nera* deserve notice, the former being made navigable down to *Peterwardein*, by an expensive canal, drawn from *Lugoſh* to *Temeſwar*.

The soil is extremely fertile. The wine is in many places excellent. It is generally of a red colour. Peach, cherry, and plum trees are very common. Large plantations of that kind skirt the villages and provide the inhabitants with their drink. The silk plantations spread almost over the whole country; they might, like many other manufactories of the bannat, be in a more flourishing state, if that great general and politician Count *Mercy d'Argenteau*, had lived to support them.

Of late there has been raised in this country a national-militia, which in the imperial and royal military state goes under the name of the *Illyrian* regiment. It is commanded by the lieutenant-colonel

colonel Baron *de Sezugafs*, knight of the *Therefian-order*, a man who has greatly ferved his country. Not fatisfied to have corrected the rough behaour of his officers, and to have habituated them to the *German* manners, he endeavours likewife to humanize his private men. He eftablifhes fchools and mafters, and the foldier is obliged to have his children fent there. If we had a calendar of political faints, Baron *Sezugafs* would fhine in it, under the title of the *Illyrian Reformer*.

The *Plajafhes* are another fort of national troops, pofted on the limits of *Tranffylvania* and the greater *Wallachia*, from *Marga* towards *Orfowa*, to put a ftop to tranfmigrations, and to prevent the efcape of the *Turkifh* and inland robbers. They are under the command of captain *Peter Vanfha*, who in the laft *Turkifh* war was *Haran-baffa*, or chief of a numerous gang of robbers, and deferved his fortune for having in the laft war faved the late emperor at *Cornua* from the imminent danger of being taken prifoner by the *Turks*.

This nation is remarkable for having produced many brave men of great defert. Captain *Ducca* for example, a man of eighty years of age, has in the late *Turkifh* war been of eminent fervice to the court; however, he never has folicited or received any preferment, happy in the confcioufnefs

of

of his honeſt ſervices, and of his maſter's grateful diſpoſition. I will in one of my letters deſcribe at large the character, the manners, and the religion of the inhabitants. At preſent I add only an abſtract of my yeſterday's tranſactions and buſineſs.

Soon in the morning I was awakened by a diſmal and frightful rattling of chains, which ſounded all along the ſtreet where I have lodgings. It was occaſioned by the malefactors, condemned to the fortifications, who, by couples chained together, went to work. I did not ſee in the ſtreets any but bleak, yellow-coloured, decayed faces, peeping and iſſuing forth from the fineſt buildings. The women, even the girls, had thick ſwoln bellies, left them by the fevers. I fancied myſelf in the realms of death, inhabited, inſtead of men, by carcaſes in fine tombs. At dinner all the gueſts, beſides me and ſome foreigners, had a fit of their fever; ſome freezing, gnaſhed their teeth; ſome burning for heat, could not aſſuage their thirſt. In the afternoon I viſited the canal which I ſpoke of before. I ſaw there ſome hundreds of bee-hives conveyed to the meadows, and to the heaths, where the bees are left for paſture during the whole ſummer. Each ſet of ſixty hives has a bee maſter to take care of them. The hives are conſtructed

of

of eleven thin pieces of deal, three inches thick, and at one end decreasing into a point. They are joined by willow or birch branches into a hollow cone, open at the bottom. Two or three inches above the ground there is a small opening, and within some crosses of wood, on which the bees suspend their work. But they behave in *Hungary* with ungrateful cruelty to these laborious insects, since, to take out the honey, they push the hive with violence against the bottom of a tub, which brings down bees, wax and honey in horrid confusion, the whole to be mashed and crushed into a sweet but disgustful mixture.

In the evening I visited the publick goal, where I saw a famous robber, who, during the last summer, had greatly annoyed the *Turks*, and by particular desire of the Grand-Signor is kept here, as they told me, till the end of the war. He is a young, well dress'd, and handsome man. He was formerly a rich merchant in *Servia*, and became a robber to revenge upon the *Turks* some violences which they had offered to him and to his family. His determined, bold, physiognomy, and his rash undertakings, in which he was very successful, raised in me the idea, that perhaps he might have proved an *Alexander*, if he had been born to attempt with greater forces, what he neither

ther dared, nor is any other perfon permitted to attempt, with fmaller ones. All this will eafily convince you that my ftay in this place cannot be agreeable to me. But the bufinefs of my companion layeth me under the neceffity to ftay fome days more. Therefore, if you be happy, remember your friend in *Pontus*.

## LETTER III.

*Temeſwar, June* 20, 1770.

THE inhabitants of Bannat are *Raizes, Wallachians,* and a fourth part *Germans.*

The *Raizes* are ſaid to be originally a *Scythian* people, in former times inhabiting *Dacia,* now called *Servia.* They call themſelves *Srbi.* Their language is a corrupt *Sclavonian* or *Illyric* dialect.

The origin of the *Wallachians* is leſs certain. They call themſelves *Romun,* a word which in their language equally ſignifies a *Roman* and a *remaining man,* and makes it doubtful whether they be remaining parts of *Roman* colonies, or of a people conquered by the *Romans.* The *Roman* medals, tombs, and other monuments, found in the mountainous parts, and near the *Danube,* are valuable evidences of their having been in former times ſubjects to the *Romans,* either in the one or in the other ſenſe. Even their language, which in greater *Wallachia (Zara more)* is ſpoken very rudely, but in *Tranſſylvania (Ardellia)* has the reputation to be ſpoken very elegantly, is a corrupt *Latin.* However, I do not conceive how ſo many

many *Italian* words, such as *rame* (copper) *mangar* (eat) and many more, that have no similarity with the *Latin*, came to be used by them. The termination of their words in general, and the conjugations after the *Italian* manner, have been mixed into the language of this nation.

Their manner of living is extremely rough and savage. They want religion, arts and sciences. Their children are from their first infancy washed every day in the open air, in warm water, and then swathed in coarse linnen or woollen cloth. The difference of the seasons and the weather makes herein no difference. From the fifth to the twelfth or fourteenth year of their age they are left with the herds and flocks to attend them; however, the girls are taught in the same time washing, baking, spinning, making needle-work, weaving, and so on. From the 14th year they are brought up and employed in husbandry. Kukuruz or maiz is their chief object of agriculture. However, they sow likewise oats, barley and corn. They distil from the fruits of trees, which they plant in great plenty, a sort of brandy, called rakie, which they are very fond of. Their meat is as simple as their dress. Bisquet of coarse grinded maiz, baked under ashes, which they call malai, some flesh, milk, cheese, beans and other vegetables, are their common food. Their dress is various;

various; but generally it consists of the following articles. The men wear long white woollen trousers, as the *Hungarians*, but wider; soles of raw skin tied about the feet instead of shoes; a shirt open on the breast; a woollen jacket or coat, tight around the waist, with long sleeves, and a fur cap or bonnet for the head. The women have long shirts down to the ancles; a brown variegated striped petticoat open on both sides, and tied with a girdle; a waistcoat or garment of coarse cloth, somewhat shorter than the shirts, and an annular bolster stuffed with hair or straw upon their head, which they cover with a woven cloth. The girls go bare headed. Their ornaments consist of ear-rings of white or yellow brass, of coloured glass, beads, pearls, glass feathers, and pieces of money fastened to a string and tied around the head and the neck. This ornament makes a ringing, so that a fine drest'd *Raize*, or *Wallachian* girl, may very often be heard sooner than seen. They marry very young; and there are married couples, the man not above fourteen, the wife even not twelve years of age. Some manual arts seem to be peculiar to them. Scarce any where you will find a cartwright, or a weaver; every *Wallachian* being a cartwright, and every woman a weaver. No woman is seen going about without some work in hand. What they bring to sale

they

they carry on their heads. If they have a child to nurfe, it is carried in the fame manner. The fpindle is fticking in their girdle, and all the way they are fpinning. All their neceffaries are worked up by themfelves. Scarce any tradefmen nor any beggars are feen among them. What can I fay to you of their religion? They confefs the non-united *Greek* religion, *Græci ritus non unitorum*. But in fact they have fcarce more religion than their domeftic animals, except repeated faftings, which almoft take up half the year, and are fo extremely fevere, that they dare not eat any meat, eggs, or milk; they fcarce have any idea of other religious duties. But in thefe faftings they are fo fcrupulous, that they do not break them, even fhould they flight every other divine or human law. A robber will never indulge himfelf contrary to this abftinence, nor lie with his own or another man's wife, for fear that God might in this cafe withdraw his blefling from his trade. What barbarifm! what humiliating ideas of the Supreme Being! The ignorance and fuperftition of the *Bonzes* cannot poffibly be above that of their Popes. Some of them are fo ignorant as to be unable to read; what can they teach the poor people? They plow and till their ground, they attend their herds like other peafants, deal in every trade as *Jews*, and get drunk at the expence of their ftupid parifhioners, who fell them their fins, and

C .fancy

fancy to be happy and to be saved if they discharge their own and their deceased relations sins at a good price. The salutary ordonnances, which her majesty the queen has published against the illicit tricks of these Popes, have proved hitherto uneffectual to rescue the people from that spirit of slavery wherewith they are subject to these spiritual masters. Her majesty's wisdom is equally eminent in protecting and propagating true religion, as in checking and extirpating superstition.*

The religious rites and ceremonies of this people favour rather of Paganism and Judaism, than of that religion which they profess. For example; no woman will attempt to kill any animal whatever it be. The bride is on her wedding day, and the day before, constantly hid under a veil. Whoever unveils her is entitled to a kiss; and, if she desire it, obliged to make her a present. The women are in the churches separated from the men. Their funerals are singular. The corpse is with dismal shrieks brought to the tomb, in which

---

\* The above instance of her majesty's maternal care for her much-beloved, faithful, and loyal Hungarian subjects, who, in the beginning of her reign, unanimously declared, *Moriamur pro Rege nostro Maria Theresia!* is, indeed, a new laurel added to the glory of Austria, by so many victories over the Turks; and of late, by so many admirable laws and establishments for the improvement of commerce, trade, and husbandry, fixed for the latest posterity.

which it is sunk down as soon as the Pope has done with his ritual. At this moment the friends and relations of the deceased raise horrid cries. They remind the deceased of his friends, parents cattle, house and houshold, and ask for what reason he left them. As no answer ensues, the grave is filled up, and a wooden cross, with a large stone placed at the head, to avoid the dead becoming a *vampyr*, or a strolling nocturnal bloodsucker. Wine is thrown upon the grave, and franckincense burnt around it, to drive away evil spirits and witches. This done they go home; bake bread of wheat flower, which to the expiation of the deceased they eat, plentifully drinking to be the better comforted themselves. The solemn shrieks, libations of wine and fumigations about the tomb continue during some days, nay even some weeks, repeated by the nearest relations. The funeral of a bridegroom is still more solemn. A pole, some fathoms long, is fixed to his tomb, and the bride hangs on it a garland, a quill, and a white handkerchief. They avoid going into our churches: If by accident the get there, they purify themselves afterwards by ablutions. To be sprinkled in our churches, or to undergo any ceremonies with consecrated water, is a matter of the greatest horror to them, because it is sprinkled about with an instrument made of pork-bristles (aspergillum)

(aspergillum.) This makes them, according to their opinion, highly impure *(sporcat,* as they call it.) Even their dresses suffering by such an accident cannot be worn again without washing. Their Popes distribute the consecrated water by a branch or nosegay of hyssop, according to the Psalm: (*Asperges me hyssopo.*) For a long while I did not understand what the *Wallachians* meant by *Frate de cruce,* or *Mangar de cruce.* At last I have learnt it. If they engage themselves in an indissoluble friendship in life and death, they put the form of a cross in the vessel or the cup from which they eat or drink; swearing everlasting fidelity. This ceremony is never to be slighted. It is generally a previous rite to robberies. The same ceremony is resorted to as the most efficacious bond; for example, if robbers release a man, by whom they apprehend to be indicted, they oblige him to silence by an oath by the cross, the salt and the bread, which they call *Giurar pe cruce, pe pita, pe sare.* Their canon law is very different from ours. Stealing and adultery are considered as trifling crimes; but violating or dishonouring a girl are great ones. No murther can be dispensed with by their popes. That dispensation is reserved to God alone. However, robberies and murthers are extremely common among this people. The reason is obvious. They have no true ideas

ideas either of God or of the soul; how should not they be wrong in their ideas of the social and political obligations of man? Any phænomenon, or effect of unknown causes, is considered by them as a miracle. They look upon a solar eclipse as a fray of the infernal dragon with the sun; for that reason, during an eclipse, a great firing is heard through the land, to frighten away the dragon, which else might conquer and devour the sun, and plunge the world into eternal darkness. The insects which in the spring creep forth from under a rock near *Columbacz* on the limits of the *Turkish* dominions, and which greatly annoy their herds, are according to their opinion vomited by the devil. The holy knight, *St. George*, is said to have cut off his head in a cavern under that rock. A *Wallachian* will never cut a spit of beech to roast his meat on. The reason is, beech yields in the spring a red sap, and the sentimental compassionate tree weeps these bloody tears according to the learned and profound observations of the *Wallachians*, because the *Turkish* bloodhounds used to cut the spits for roasting Christians from beechwood. No capital punishment is in greater abhorrence amongst the *Wallachians* than that of the rope. The pale and wheel seem preferable to it. But why? A rope ties the neck and forces the soul out downwards. They call that a most disgustful impure defilement of the soul, and I call their singular nicety on that account true psychological materialism.

Superstition being the daughter of folly, you may easily guess by the above instances how remarkably ignorant they are. Ask an old *Wallachian* what age he is? He will answer at the siege of *Belgrad* or *Tem-swar*, at the conclusion of the peace, or when that prince died, or that metropolitan was elected, I attended the swine or the sheep, I went into the field, I married, and so on; and then you may cast up his age. They are not generally acquainted with the value of the current money. Even its denominations are not taken from their own language. A dollar, or thirty groshes, is called *leu*; a florin, *florint*; a half florin, *dult*; five groshes, a piece of their currency, is called *Strimbe*; half a dollar, *tri strimbi*. They have scarce any knowledge of the measure of liquids. The contents of a vessel is estimated according to the weight of the liquid contained in it. Their weight is the *occa*, a *Turkish* weight, answering to our two pounds and an half. One *occa* contains four *litre*, one *litra* an hundred *drams*.

The difference in the character of the *Raizes* and *Wallachians* is nearly as follows:

The *Raize* is fierce, proud, bold, cunning, a friend

friend of trade, fit to be a foldier. His Popes lefs ignorant than thofe of the *Wallachians*.

The *Wallachian* has no idea of haughtinefs, is a better hufbandman, a friend of eafe, and abhoring military life. They agree in being born robbers and flaves to their popes and national magiftrates. The *Greek* alphabet is ufed by both thefe nations, but they give to feveral letters a different fignification. However imperfect this fketch may be, it will do to give you fome idea of a nation, which, as far I know, is ftill deftitute of an hiftorian, to acquaint the reft of *Europe* of its origin, cuftoms and manners. To-morrow we fet out, and by the next poft I hope to fend you a letter, which, containing objects, nearer to our tafte, will prove more entertaining.

## LETTER IV.

*Oravitza, June* 23, 1770.

IT is not worth while to transcribe to you, my dearest friend, the *Wallachian* names of the insignificant villages, which I passed between *Temeswar* and *Oravitza*. The plain which from *Temeswar* stretches to *Theissa*, continued six stages more, or for twelve *German* miles to *Oravitza*.

Some hours before I reached this place, I saw to the left of the road some hills, which consisted of shivery, micaceous clay, and formed the promontory. Insensibly we ascended these hills, and reached at last the valley wherein the place is situated, from which I write to you these lines.

Here the argillaceous slate disappeared under the limestone, which hereabouts covers the surface. As soon as I arrived I called on Mr. *Delius*, whom, till now, I knew only by the reputation of his solid learning. But I was disappointed, the arrival of Baron *Hegengarthen*, commissioned to examine and to improve the mines in the Bannat, did not leave him the requisite leasure to favour me with his remarks on the nature of these mountains, which as an exact observer he ought to be

very

very well acquainted with. However, I have found another fkilful miner, who yefterday evening gave me fome particulars of the general divifion and other circumftances of the Bannat-mines. This has furnifhed me with the materials of this letter.

A line drawn from north to fouth through *Temefwar*, divides the whole country into two parts; that to the eaft is generally mountainous, and here you are only to fearch for mines. As often as I fpeak to you of their fituation, it will conftantly be in relation to the chief place, which is *Temefwar*. The mines now working in the Bannat are to the eaft, the ironworks *Bogfhan*, properly *Paffioven*; clofe to which is the new eftablifhed iron-works *Refhiza*. Thence fomewhat more to the fouth, are the copper mines *Dognafka*, farther off, *Oravitza*, *Safka*, and entirely to the fouth, *Bofniak*, or *New-Moldava*. In the plains bounded by the mountains of *Oravitza*, *Safka*, *Bofniak*, and thofe which run along the *Danube* and make the eafterly limits, they wafh Gold from the *Nera*, and *Menifh-rivers*, nay every where from the ground which is adjacent to them. In former times private companies carried on thefe wafh-works in the *Karanfebefe-Diftrict*, at *Konigfeg*, and in other places; fome are ftill going on.

All

All these mines are divided into four mountain-districts, which are called *Berg-Aemter*; such are that of *Bogshan*, to which *Reshiza* is to belong for the future; that of *Oravitza*, that of *Dognazka*, and that of *Saska*, to which is incorporated the market-town *Moldova*. They are under a direction, in which the president of the country generally presides. In former times it was at *Temeswar*; but for the future the president, a counsellor of the reports, and a secretary of this direction, are the only persons ordered to reside there. The other members, and whoever belongs to the chancery and the accounts are to reside at *Oravitza*. The mines near *Grofs-Wardein*, and the bailiwick *Refzbania*, in *Upper Hungary*, are under the same direction.

LETTER

## LETTER V.

*Oravitza, June 26, 1770.*

ORAVITZA, as I have told you, already is the chief place of mines in the Bannat. The mines of its dependency were worked by the *Turks* as long as it was under their dominion; but with lefs profit than at prefent. After the reftoration, the old mines were drained at the expence of the imperial treafury, and fome new ones fet at work; but all thefe mines, the royal galleries (ftolln) excepted, were left afterwards by grant to feveral private companies, under feveral conditions and refervations, which being merely œconomical, the tranflator fuppofes to be unentertaining, and ufelefs to the *Englifh* public.

## LETTER VI.

*Oravitza, June* 27, 1770.

THE valley wherein *Oravitza* is situated, is bounded to the south by the *Wadarna, Csiklova* and *Temese* mountains; but to the north by those of *Coshowiz, Dilfa,* and *Cornudilfa.* The mountains are here, as generally in the Bannat, gently ascending, and grown over with beech, birch, fir, ash and oak. Their rocks are argillaceous, mixed with sherl, mica, and feldspath; and this is covered either with argillaceous micaceous slate, or with a fine arenaceous or lime-stone. Between these last sorts of stones occur the copper-fissures *(Klufte)* which really deserve this denomination rather than that of veins, since they have neither a constant dipping nor a constant run. There has not yet been discovered at *Oravitza* any fissure, running or dipping above fifty fathom *(Klafter.)*

I have examined the *Coshowiz* mountains, and found in them the following mines: *Rocchus, Erasmus, Jacobus, Benedictus, Gabriel, Paulus, Genovefa, Philippus, Maria, Maria Theresia,* or the *Goldshurf, Ladislai, Pyrite-mine,* and the *Kies-stock,* where a hundred weight (1. centner) yields seventy
pounds

pounds of stone or lech. These mines are for the most part drained by a gallery (erb-stolln) which is driven in the field above 229 fathoms, and runs 19 fathoms below them. *Rocchus* is the richest; and on this account the chief gallery is driven to the south. Several drifts of smaller galleries serve to search and to work out the smaller fissures. The hanging side is limestone, the hading side slate. So it is likewise in the other mines, in these and the *Wadarna, Csiklova,* and *Temese* mountains, with the single difference, that according to the different situation of the mines, the limestone is either on the hanging or on the hading side, and that the sand-stone is often in the place of the slate on the opposite side of the fissures. The *Gang-rock* (that is to say the rock which fills the fissures) is for the greater part either calcareous or selenitic. The purer and the more sparry it is, the richer the ores contained in it.

I was very glad to be here convinced by my own experience of what Mr. *Delius,* in a *Vienna Magazine,* has published on the origin of the metallic fissures, and laid down as axiomatical rules for the mines in the Bannat; that is to say, that the metallic fissures are never to be found in the rocks, but between two different sorts of rocks. Full of this opinion I examined the *Cornudilfa* mountains, where in the *Trinity* mine I was assured of the hanging sides being lime-stone, and of the

hading

hading sides being horn-stone. However, I took with me samples of these stones, as I used to do, and trying them with steel and aquafortis, I found that they are common grained lime-stone, and that the miners had denominated one species horn-stone, for its being finer grained and harder. This erroneous denomination of the miners may propably have led Mr. *Delius* to the above erroneous assertion. The same is observed in the other mines of the *Cornudilfa* mountains which are entirely calcareous; and in those of the *Dilfa* mountains, where generally the finer lime-stone goes under the false denomination of horn-stone. The gang or vein rock, in these *Cornudilfa* and *Dilfa* mines, is either granulated white or yellow gypsum, or selenitic spar, which by a light warming gets a phosphorescence in the dark. The fissures of the *Cornudilfa*-mountains have a more even direction than in the others; but in *S. Servatius*, a mine working in these mountains, all the fissures are cut off by a brown argillaceous vein.

The *Wadarna*, *Csiklova*, and *Temesc* mountains and fissures agree in general with the *Coshowiz* mountains described before. One of the *Wadarna* mines, called St. *Paul's conversion*, yields some silver and arsenical-copper ore. I would not tire you with the list of the many mines which in all these mountains are working.

The

The common ores dug at *Oravitza* are a pale, yellow, copper pyrites; *pyrites cupri pallide flavus Cronstedts*, §. 198. Blackish grey, copper pyrites; *pyrites cupri griseus, ibid.* The last is often variegated in the surface. A species of pyrites, penetrated and incrustated with a brown copper mulm, is called broth ore *(Brüherz.)* It is found in *Trinity* mine, in the *Cornudilfa* mountain. This ochraceous ore is probably owing to a decomposition of the copper pyrites. The white arsenical copper ore, described by *Cronstedts*, §. 199, is common in the *Wadarna* mine, called St. *Paul's conversion*; but less white than that which breaks in *Herrn-grund*, in *Lower Hungary*.

In the same *Wadarna* mountains they found, twenty years ago, in St. *Anthony's* pit, beautiful *malachit ore. Ochra, veneris, calciformis, impura, indurata.* I could get no specimens of it; but they brought me from *Trinity* mine a fine *chrystalised azure copper.* The crystals are oblong, quadrangular, truncated. But to my still greater satisfaction, I got here many pieces of *red copper mulm (Ochra, Cyprii, Linn)* either dissolved in loose dust, or indurated and staining the fingers. *Cronstedts* has not described this species, unless it be that which §. 196, or 194, n. 4. he notices as found at *Sunnerskog*, or *Ostanberg*. It is likewise unnoticed by other mineralogists. Among several samples, a loose *cinnabar red ocher*, which girts a

piece

piece of native copper, is highly remarkable. Its colour so high, as to mislead even the most intelligent connoisseur. If *Cronstedts* assertion be true, that copper, by the loss of its phlogiston, may be changed into copper-glass, one might guess by the richest of this ocher, which is 54 pounds per hundred weight, that it is a solution of copper-glass. But it is found in so large lumps, that this opinion is scarce admissible. The sameness of the colour has caused this ocher to be called *tile-ore*.

The following particulars of the pay and labour of the miners are left out, as uninteresting to an *English* reader.

LETTER

## LETTER. VII.

*Saſka, June,* 30. 1770.

FOUR hours journey from *Oravitza* to the ſouth is *Saſka,* where I arrived yeſterday before night, under a convoy of ſome *Huzzars* and *Wallachians.* The country between theſe two places is in this ſeaſon extremely fine, and offers a continual variety of orchards, cultivated fields, meadows, plains and hills. The road runs all this way over glimmery argillaceous ſlate, which is now and then interrupted by ſome rocks of a grey argillaceous ſtone, mixed either with mica or ſherl, with mica and feldſpath, or with rocks of gneiſs. *Saſka* is ſituated in a valley ſurrounded with calcareous hills, ſuperincumbent on ſlate, whoſe diſſolved parts are carried by the rain water into the valley, and incruſtate there the roots and moſſes. The copper fiſſures, or veins hereabout, run between this grey limeſtone, and a margaceous rock mixed with baſalt grains, the former being generally on the hanging and the latter on the hading ſide. I may be wrong perhaps, but I do imagine the origin of this hading ſide to have been as follows: The argillaceous grey ſtone, mixed with mica, baſalt and little quarz and feldſpath-grains,

D            which

which I shall call henceforth metallic rock (Saxum metalliferum) because the nobler veins in *Lower-Hungary* constantly cross it, and because the *Saxum metalliferum Linnæi* has so great an affinity with it; this rock I say may perhaps have been incoherent still, or less indurated when it was covered with limestone, and by that accident have been changed into its present margaceous nature. Any subsequent alteration or commotion, changing their former horizontal position into a dipping or oblique one, may easily have separated, and split fissures along their skirts, which are now filled with metallic masses between the calcareous hanging and the margaceous hading side. These mines were taken up again after the restoration of the Bannat, about the year 1746. They worked first on some copper veins, lying open to the day. Then the *Wallachians*, who had been searching after mines, discovered some old pits and overgrown large bings, which proved that in former times miners had been working there. I have been shown myself in the higher mountains a great many pretty pure copper and lead flags, which evidence old parting-furnaces, though thereabout there does not appear any water sufficient to work the requisite bellows. Might not perhaps the ancients have trod their bellows, or worked them with engines? Might not they perhaps have smelted their ores in smaller furnaces with hand-bellows, in the same manner

manner as the *Finnlanders* and *Russian* peasants? At present the number of the several pits and drifts on small often inconsiderable copper or lead veins is astonishing. The chief are in *the Promontory* now *Nicolas*, *Theresia*, *Nepomucenus* and *Philip Jacob* mine. This last I have examined. It is one of the richest works at *Saska*. The gang-rock is as generally at *Saska* calcareous or selenitic spar, which very seldom alternates with quarz. In the *anterior middle mountains* is *S. Mary*, and other insignificant stock works, if small nests or lumps deserve that name. In the *upper middle mountains* are holes or quarries, whence they dig out from between the vegetable mould and the lower limestone a brown irony earth, which yields from two to six pounds of copper per hundred weight. The most considerable pit of that kind is in the higher mountains, called *Maria felsen*. It may have three or four fathom diameter on an equal depth.

Mr. *Delius*, in the above quoted Treatise, gives the following conjectures on the origin of this copper mulm:

" Water is endowed with the quality of de-
" stroying the form of metals, and restoring them
" to another form; I mean only to speak of the
" mixture and external form of their masses, not
" of their constituent metallic particles, which are
" eternally permanent. This happens very com-
" monly to the copper ores, and generally to those
" which

" which are pyritical ; they are subject more than
" any others to be dissolved by water into vitriol.
" Such a transformation of ores nature has produ-
" ced in the mines of the Bannat near *Saska*, where
" a whole mountain has been subject to it. The
" copper is not mineralized, but it appears as a
" metallic dust in a brownish earth, if properly
" washed. This earth did not exist from the begin-
" ning in that form ; but it was rather a copper-
" pyrite, which by the waters has been dissolved.
" The sulphurous acid went off washed away by
" the water ; but ocher and the unmetallic earth,
" which are the constituent parts of the copper py-
" rites, remained, retaining the native copper par-
" ticles as a filter. This formed the *Saska* copper
" mulm." Mr. *Delius* supports this opinion by
the copper-pyrites still found in the mulm, and
yet unaffected by the dissolution ; but however
probable it be, he has neglected a circumstance
which I confess myself to be scarce able to answer
for. The brown copper mulm under consideration is immediately under the vegetable mould,
and its hading side is limestone. Is there any
probability that these pyrites have been before
their dissolution without any roof or hanging
side ? If that be the case, the *Romans*, which,
according to Mr. *Delius*, have worked in these
parts under *Trajan* and his successors, might have
driven expensive shafts and galleries, whose cop-

per

per remains are still found in these mines? But they had a more easy way of getting copper. They wanted only to get the ore immediately, and by the least trouble. *

Might we not rather conjecture that these fissures, when the ancients worked in these parts, had then their own hanging and hading side together, with a different position; and that after a previous earthquake it has been changed so as to deprive these fissures of their former roof and hanging side, and to expose the ores to destruction? An able miner, used to observe nature, might perhaps rectify these conjectures, which I scarce am bold enough to venture, as having had no leisure for proper examination.

*Bona Spes, Anna Rosina, Maria Snow, Mary's Visitation* and *Bonifacuis*, in the higher mountains, are equally remarkable mines, on account of their beautiful ores; and *Saska* is perhaps that place, which has supplied my collection with the richest crop of mineral curiosities. All the different species of copper ores, that of the *Mansfield* copper slate

---

\* The translator sees not the least consequence in this whole argumentation, as these pyrites might have been with other rubbish washed and accumulated on the limestone-ground, and the ignorance or neglect of the *Romans* cannot be fairly alledged against the hypothesis of Mr. *Delius*, since it proves too much or nothing.

excepted, and many more new and unknown ones, are dug hereabout in great plenty. In *S. Urban* I found native copper with a polished splendent surface, sticking to a matrix of clayish sandstone and quarz; and in the *New-Elias* I got native branchy and dendritical copper in white indurated clay. Native copper in loose brown copper mulm from the before described pits in the higher mountains, and in green and blueish copper ocher from *Mary's Snow*, are not unfrequent. I was presented with a sample of native woven copper, by its texture greatly resembling the woven silver, from *Johan George-Stads* in *Saxony*. This species is found in *Bena Spes*, in quartzous ganglock, mixed with greenish lithomarga; grey copper glass, *Cuprum Sulphur mineralisatum solidum textura indeterminata Cronstedt*, §. 197, is found in *Philippi Jacobi* pit. It is malleable, and of a compact texture. They call it here *lech ore*. It breaks in scaly grey limestone; yields from sixty-three to seventy pounds of copper, and moulders by dissolution into a blackgrey dust. Red copper glass of an undetermined figure, *Minera cupri calciformis pura & indurata colore rubro Cronstedt*, §. 195, found in *Maria Brunn*, in a white gypsum tinged by verdegrease. In the same place it breaks in a fibrous verdegrease, which makes it very beautiful to the eye. Mr. *Delius* presented me with such a crystallized copper glass, which

consists

consists of many accumulated red transparent triangular crystals. These, and a variety of octangular crystals, are found in *S. Urban* and *Mary's* visitation sticking to an undescribed copper ore. It is a brown red fine grained jasperlike stone, striking fire with steel. I might by Mr. *Cronstedt*'s example, who calls our *Hungarian* zinopel jaspis martialis (Minerol. §. 65.) name it a copper jasper. It contains, separated from the richer crystals, from thirteen to nineteen pounds of copper. Some pieces mouldered into a red copper ocher, and containing in the middle only a remaining kernel of this red jasper, convinced me that the *tile-ores*, which are dug in the same mine, and which I have described at *Oravitza*, owe their origin and their riches to this jasperlike ore, and its copper glass crystals. Among a variety of verdegrease, *Ochra cupri viridis, viride montanum*, which is here very common, I received fine fibrous glossy copper green *Aerugo Linnaæ*. The fibres are for the most part concentric, pointed below but two or three inches, large and flat at the top. They call it *satin-ore* (Atlas-ore.) There is an innumerable variety of *malachites*, in thin flat plates, but knotty, in concentric coats, in thin undulated lamellæ and scales; its colours from the lightest to dark green in every sort of shade. The *Barmaster* in this place improved my collection by a sample from *Reczbania*, a copper

work

work in *Hungary*, under the direction of the Bannat. It is an indurated fibrous verdegreafe (*Ærugo Linnæi*) which after the tranfmutation into malachite has preferved its original concentric fibres. Indurated *copper azure (Cærulcum montanum induratum. Cronftedt.* §. 194.) and cryftallized azure, in gloffy femi-tranfparent polyedrous cryftals, offered to me in *Urban* and *Maria Shutz* mine. I gathered here for my mineralogical friend a good ftock of the brown and grey copper mulm. Such an indurated mulm from *Philippi Jacobi*, and other mines, mixed with fome phlogifton, fmooth and gloffy where broken, is on account of the likenefs called *pitch-ore*. It feldom yields above feven or eight pounds of copper; but being commonly mixed with verdegreafe, azure, cryftallized red copper glafs and native copper, it is generally ranked among the richeft ores of *Safka*. Befides the *fallow copper ore*, *pyrites cupri grifeus. Cronftedt.* §. 198. which they call here *white-ore*, any other fort of copper pyrites are found in thefe mines. In the upper middle mountains they find in fome lead pits a light brown *lead ocher*, which is often mixed with white irregular fpar-chryftals. The above *pitch ores* are commonly covered with blue columnar hexagonal or polyedrous gloffy cryftals, truncated on both ends. They never contain any copper, and are but *blue fherl cryftallifations*. Mr. *Dembfher*, a very intelligent

intelligent affayer at *Moldova*, has affured me, that for a long while he had, without any fuccefs, affayed thefe ores as copper ores, till at laft he had found in *Lehman*'s preface to *Marggraf*'s works, that now and then handfome blue chryftals had offered to him entirely deftitute of copper, but containing plenty of iron. The only remarkable ftones, which, befides the different rocks I have found here, are a white tranfparent calcareous fpar cryftallifation, confifting of columnar hexagonal cryftals, with three large and three fmaller oppofite fides and a triangular point; a dodecaedrical cryftallifation, compofed by pentagonal faces, drawn in *Linneus's Amocnitatibus*. tom. I. fig. 25; and a pyramidal, triangular, tranfparent felenit-cryftallifation. As foon as I return I will divide my collection with you, knowing very well that they have raifed your curiofity.

LETTER

## LETTER VIII.

*New Moldova, July* 1, 1770.

THE daily examples of the ill use, which travellers in the Bannat, especially in these parts, are exposed to from the numerous gangs of robbers, had almost brought me to the resolution to give up my journey to this last boundary, which is separated from the *Turkish* dominions by the *Danube*. But I heard that these fine gentlemen vent their ill humour rather against their countrymen, which have the misfortune to fall in with them, than against any *German*, which is said to happen but very seldom. This circumstance, and my recollecting the chiefs of the robbers having sent word to the aulic commissioner, Baron de *Hegengarthen*, that he and his men might in safety, and without any convoy, travel where he pleased, gave me resolution to attempt this excursion towards the east. Twelve mine-officers on horseback, and some common miners armed with guns, went with me. As soon as we had ascended the higher mountains of *Saska*, I observed that gneiss, now and then cap'd with common clay-schistus and limestone, covered the whole

whole country. This continued to *Moldova*. Some small copper veins baffet out from the schistus. But the skilful miners do not work them, because they dip only some feet in the slate and then strike dead or disappear entirely. Perhaps after a long series of years they will dissolve into copper mulm as at *Saska*, and then be got easily by posterity. After two hours ride we alighted at a copper furnace in the midst of a thick forest. The mine-officers from *Moldova*, and about thirty armed miners, had expected me there and joined our caravan, which now ressembled a little army. I was agreeably surprized to meet here with my old college acquaintance, Mr. *Dembsher*, assayer and engineer at *Moldova*. This young man, possessed of all the theoretical and practical science of miners, of much learning and good taste, has for several years, by our continual correspondence, prepared me for a journey to the Bannat, and enabled me to make and to justify in a short time all the observations which I have given and shall continue to give you. His conversation, and the merry chearfulness of my convoy, diverted me so much, that I thought of no danger in the thick woods which we crossed to *New Moldova*. As soon as we arrived I visited the town *Moldova*, at the foot of the mountains on the *Danube*, to see some robbers who had been taken by a party of soldiers. They had brought along with them the head of a young man, who

had

had bravely fought againſt them, and preferred death to chains. In the evening I returned to *New-Moldova*, or as they call it, *Boſniak*. The fine views from the hills were extremely pleaſing. I ſaw from thence a large tract of country far in the *Turkiſh* dominions; but I did not ſee without concern the hills, which concealed from my eyes the former rich copper-works, near *Maidenbeck* in *Servia*.

To day I viſited the mines hereabout. They are divided into three diſtricts, that of *Benedicts*, *Florimund's* and *Andreas*. In the firſt are *S. Barbara*, *Trinity*, *Nepomucenus*, *God's hope*, and *fourteen Nothelfer*; in the ſecond, *Joſeph*, *Thereſia*, *Archdutcheſs Mariana*, *Pelagia*, *Maria good Rath*; and in the third, *Andrews*, *Peter* and *Paul*, *Anton* from *Padua*, *Hilarius*, *Thomas* and *Helen*. They are all working, and yield fine copper ores from veins running in almoſt every direction. *Maria Thereſia* yields lead. The hading ſide of theſe veins is grey clay-ſchiſtus; the hanging ſide is limeſtone; both ſuperincumbent on gneiſs. Theſe mines ſeem to have been worked in times of old; ſince the miners ſcarce have reached till now any ſound or new field, and get their ores only in the old man. The ancients have indeed left ſtupendous works in the *Beſedine* mountains, which are not worked at preſent. They have formed works with chiſſels and hammers in rocks, which we hardly

hardly conquer by blasting. In some parts the walls are so flat and even, that they resemble rather stonecutters than miners work; where they met with loose crushed rocks, they left tremendous caverns. It is astonishing that the most ancient works are generally driven in the soundest rock. Whether they may be ascribed to the *Romans* cannot possibly be ascertained. The construction of these old galleries and drifts has nothing particular; it agrees with what you have seen in the *Trinity Erb Stolln* at *Shemniz*. The doors are either cut in solid rock or lined and fastened by uncemented masonry; their figure eliptical. They work here as at *Saska* on fissures, which are inconsiderable. The ores found in this place give the most malleable and tough copper in the Bannat. For this reason, and to encourage the working of the *Moldova* mountains, the imperial direction pays for the *Moldova* copper four florins extraordinary. Almost every sort of copper ore which I have mentioned, from *Oravitza* and *Saska*, are found here. *Native copper* breaks in *God's hope* in different forms. It sticks commonly to quartz. If found on black grey copper pyrites, it moulders in open air into a calx resembling pulverized tiles, but whitening still more and more. In this state of dissolution it scarce yields any copper at all. The native copper from *Johan Nepomucenus*, and *Barbara stoln*,

are

are of the same nature. Red copper-calx is found in *Archduchefs Mariana* in a matrix of asbestus, which contains likewise now and then some copper pyrites. In *Hilarius* I got some fine red copper glass crystals, and from the old bings of the *Besedine* mountains such crystals, and lead glance, which contain some silver.

My friend *Dembsher* assisted me to get here a large stock of pitch, broth and clay ores, or of whatever other sort and denomination of richer ores. With this booty I return tomorrow to *Oravitza*, to pay my last respects to Baron *Hegengarthen*, and to pursue by the road of *Dognazka* my way to *Transsylvania*.

LETTER

## LETTER. IX.

*Dognazka, July* 5, 1770.

THE day before yesterday I took my farewel of Baron *Hegengarthen*, and arrived here after five hours ride. Clay-schistus, mixed with mica, cap'd the lower granite, which now and then peeped from under ground, all the way long from *Oravitza* to this place. The mountains, which are working at *Dognazka*, are middle mountains, which rise from the plains near *Werschez*, and run eastward to *Transsylvania*. The chief ridge of these mountains is granite, covered by gneiss, clay, sand and lime. The only constant vein (gang) in the Bannat is here at *Dognazka*. Its run and dipping is constant for a great while. It is situated in *John's mountains*, and consists of a lead and silver vein. They have chased it already in its run from west to east 1500 fathom. It dips from south to north. Before the last *Turkish* war they got here a good deal of silver. The following different pits, *Mary Christina, John* and *George, Susanna, Nepomuck, Barbara, Samuel, Mercy, Sweti Theodor,* and the *Herbestine* stoln are at present working upon it. This vein running along

the

foot of the higher incumbent lime and slate-hills, the mines are greatly exposed from water. For the greatest part of the year they are under water; and though in *Maria Christina* they have of late built a horse-engine, in hopes to drain this mine, I am apprehensive it will fall short of expectation. This very circumstance hindered me to examine these mines myself, which I should have been the more inclined to do, as I cannot believe that the hanging and hading side of these veins in greater depth, but consisting of lime and slate, or of hornstone and argillaceous slate. However, that is the assertion of the mine-officers. I have examined the rubbish of these mines, and true it is, that it consisted of indurated shivery clay and limestone; but as this may be supposed to have been drawn only from the upper drifts, I am still of opinion, that in a greater depth gneiss or sherl-mixed argillaceous rock, Saxum metalliferum might be found. It is highly improbable that a constant quick vein should have such an uninterrupted run in rocks so accidental as these superincumbent clay, slate and limestone hills; and I have found in an old account of the *Bannat* mines, from the year 1748, that in the then new imperial stoln or gallery, after it was driven through the slate, they reached a very hard rock, which made the work go on very slowly; and that in the old *Josephi-gallery* they met with a

rough

rough hard rock, which determined the proprietors to drop it entirely. Every enquiry was unsuccefsful, since the mine-officers in these parts do not know any rocks but limestone and slate, and since accustomed to search after and to find their ores between these rocks, they neglect to observe any other sort. They have this fault in common with the mine-officers in the imperial states in general.

Ask them the nature of their mountains? and I am sure they will give you so indifferent a description, that you cannot make any thing of it, but that they never troubled their heads with such observations. However, the surest rules of a rational working of mines entirely depend on this neglected science of the mountains, their strata and their varieties.

I might alledge to you many examples of the ill consequences of this neglect. I examined a mine in *Hungary*, which in former times had yielded a rich overplus from a pyritical vein, containing gold, and crossing an argillaceous rock adjacent to granite. This vein was cut off by the granite, and as the same seems to have happened to a great depth the ancients gave it up. Of late they resolved to take it up again. They followed the line of the compass in which the vein was known to run, and drove a long drift through the granite to meet again with the lost vein; but without any success. The same would happen if they should search after it in the direction of its dipping; and all these pains and expences would have been saved if previously they had consulted and considered the nature of the rocks. After this occasional digression I return to *Dognazka*. Besides the before mentioned mines in *John's mountains*, there are several lead and copper fissures working in the *Wolfgang Dilsa* and *Morawiz* mountains; such as *Mary victory*, *Christoph*, *Traugot*, *Pancratius*, *New Gluckauf*, *Erasmus* in the former; *Rochus*, *Fabianus* and *Theresia* in the *Dilsa* mountains; and *Francifcus*, *Peter* and *Paul*, *Johanna*, *John Baptist*, *Trinity*, *Maria Litchtmass Paul* and *Simon Judas* in the *Morawiz* mountains.

*Simon Judas* is perhaps the most considerable copper mine ever discovered in Europe. After many insignificant searches on the surface, a company of adventurers united in the year 1740, and in hopes of some silver ores, pursuing the upper gallery in a dead fissure, drove many drifts in the field; but the adventurers gave it up, and a single remaining tenant, after being ruined, and having attempted a drift to the east, discovered a rich copper fissure, on which at last they sunk a shaft. The copper fissures crossing and uniting here from every side form a stockwork, as they call it here, though it be different from the stockworks after the *Saxonian* principles, which are entirely
independant

of the rest of the mountain, and are said never to have any hanging or hading side. An unhappy avarice prevailed then on the associates, to encourage the finding of copper by prizes. The barmaster was allowed five, and the furnace master three groshes per hundred weight. This caused the barmasters to work as farmers, and to consider only their present advantage, without any regard what was to become of the mine in future times. Accordingly immense quantities were taken from this stock of rich ores, and tremendous caverns produced, which, unsupported, threatened inevitable ruin. But Count *Gotlieb Stampher*, at *Shemniz*, was at last commissioned to examine and to prevent this bad practice; and Mr. *Delius*, then barmaster at *Dognazka*, ordained, that from the bottom of the stock upwards to the ninth level or gallery, the whole cavern should be filled up with deaf rocks, except some small doors and pits, left open for procuring the ore remaining in the depth. By this means the danger was prevented, or rather lessened. This remarkable work I examined yesterday. The rocks which surround t are Saxum metalliferum mixed with some lime. The gallery at the ninth level has been driven through it. The stock itself, at least that part which is above this ninth level, has a hanging side of a scaly white limestone, and a hading of slate; but the whole stock, or rather all these united veins, are incumbent on gneiss.

This

This gneiss-ground has been explored for copper fissures by shafts and drifts; but finding that there was no chance of success, the inferior part of the stock has been filled up, as I told you before, and this hindered me to see whether it does not consist of Saxum metalliferum, and whether this variety of rock has not produced the irregular dipping of the vein, which at present is ascribed to the large *Francisci* vein crossing in from the hanging side. At my entrance in the stockwork, I was greatly surprized by a magnificent view, which however, at second thought, I found equally tremendous. The whole and wide cavity of the mine was illumined with a vast number of tapers, and the workmen stood or appeared hanging on the projecting stripes, or soles of rich various coloured copper-ore. The form of this stockwork is oval; its uppermost or first level has a breadth of three or four fathom; but it increases so much, that on the ninth level it has twenty-six fathom length and twenty fathom width. From this level it decreases in the same proportion towards the under part. I have told you already that this copper stock, or copper belly, is meerly produced by the coincidence of several veins, on which account it cannot be compared to other stockworks, as that for example at *Geier* in *Saxony*; but it has likewise a visible run from east to west, and a dipping, which from the first level to the ninth goes from

south

south to north, and thence to the sixteenth level, in an opposite direction, that is to say, from north to south. The depth of the whole stock is forty fathom, and the *Joseph-Shaft* is sunk into it. The rubbish and ores are drawn out by horses. So are the waters, which from the deepest sole are pumped up to the ninth level, where they are carried of by a gallery. The annual dividend of this work is at present greatly decreasing, so are the ores; and probably in ten or twelve years time the works will be at an end, since the many crossing veins, by their opposite directions, strike dead the rich fissures, concentrated in this spot of ground. The searching drifts on those cross veins give no hopes. Nevertheless they get still every month 7 tons and a half of copper. The ores are lying in so close a mass together, that scarce any deaf rock is to be seen or dug out. For this reason the supports of the roof, and the stairs to the first, second, third and ninth level, which are still found, are cut in the finest variegated copper pyrites. The gang or vein rocks, which now and then offer, are a fine white and scaly limestone, calcareous spar, white achate with red and black spots, and yellow or black granulated garnet. (*Granatus figurae incertae particulis granulatis. Cronstedt.* §. 69.) It is remarkable that, about an hundred fathom distant from the hading side of this stockwork, the *Paul*'s lead mine, and in about a similar distance from its hanging side, an iron mine is working. The ore of the latter is sent to *Bogsham*. As probably some fissures of these veins are crossing over to the adjacent stockwork, there occurs not only in the hading side of *Simon Judas* leadglance in copper ore, and yellow crystallized garnets, which in *Paul's* lead mine are extremely frequent in and next to its lead ores, but in the hanging side the ores are striped and penetrated with ferruginous ocher. *Mary Victory's* mine in the *wolfgang* mountains has been but of late begun working. It is in metallic rock, *Saxum metalliferum*, which the miners hereabout call sandstone. The vein or gangrock is a fine dissolved white mica or glimmer, mixed with stone, and blended with copper pyrites. There is hope of good success. In *John Baptists* mine, in the *Moraviza* mountains, breaks white alabaster, girt with limestone and slate, and containing copper pyrites. The vitriolic acid of the pyrites might perhaps have changed the former alcaline gangrock into a gypsous substance. The *Isidore* mine has been dropt some years ago, for not answering the expectation. It seemed to me however very remarkable, as being for a long way covered with a brownish-yellow asbestus, containing ironglimmer and black iron garnets. This asbestus introduces itself into the copper fissure, and is the matrix of the copper pyrites. I made at *Dognazka* a large collection

of

of scarce ores; and among the before described species I got the following samples:

Native lamellated gold in a brown iron clay from *Fabianus*. In this mine it is often found in lumps included in the copper vein.

Native copper from the same place, in large heavy lumps, which might be considered as smelted, if the red copper glass crystals, that surround them on every side, did not prove it to be produced by nature.

Native copper in brown iron ocher from *Simon* and *Judas* stockwork.

Grey crystallized copper glass ore. The crystals polyedrous sticking on quarz, from the same place

Grey variegated copper pyrites from *Simon Judas*, called copper glass, on account of its content of sixty to seventy pounds of copper, glossy on the fractures; differing from other copper pyrites by its red and blue colours not being superficial but penetrating its whole substance.

Red crystallized copper glass.. The crystals oblong prisms, truncated on both ends. From *Paul*'s mine. Scarce; found in brown copper ocher.

Red copper mulm (tile ore) girt with a coat of verdegrease, which seems to be produced by an acid solution of the copper mulm. From *Mary Lichtmass* at *Dognazka*.

Grey

Grey copper-pyrites) *Cronſtedt.* §. 198.) cryſtalliſed. The cryſtals have ten faces. From *Simon Judas.*

Grey and yellow mixed ſcaly copper pyrites; greatly reſſembling our ſcaly cobalt (ſherben-cobolt) however different on account of its yellow colour, from the ſame place.

Yellow and black undeterminate garnets, in large pieces, from *Paul*'s mine. They call them yellow or black hornſtone.

Yellow garnets of eighteen and thirty ſix points; often of the bigneſs of a pigeon egg. The miners call them yellow blend. From the ſame place.

# LETTER. X.

*Lugos*, July 7th, 1770

FOR this letter you are indebted to the neglect of the poſtmaſter. I ordered the horſes at four in the morning; but he cannot procure me any before ten. For ſome hours I have been in this market town, viſiting ſome of my acquaintances, who commonly in this ſeaſon flock hither to a healthier climate, from the raging fevers at *Temeſwar*. I owe you ſtill a remark on the ſmelting in the *Bannat*; and here you have it: The ſmelting and refining of the copper at *Oravitza* is nearly the ſame which you have ſeen in the *Lower Hungarian* works, and is done in four different ſmelting places, called the *Franciſcus Mercy*, *Thereſia*, and *Saiger-hutte*. Great care is taken in rejecting the refractory ores.

Two tons of ore, twenty-four of pyrites and twelve carts of copper-ſlags are commonly put together in the firſt ſmelting. If the ores be remarkably ſulphurous the quantity of pyrites is leſſened; ſo the quantity of ſlags if they be mild. In twenty four hours time the buſineſs is done. The whole gives

gives about three or four hundred weight of copper.

Seven tons and half of *raw-stone (Rohstein)* produced by the first operation, make a roast.

The *black copper*, procured by roasting, is refined on a smaller hearth, and in smaller quantities of about four or five hundreds.

All expences cast up, a hundred weight costs the proprietors from nine to eleven florins.

The parting furnaces *(Sayger-hutte)* are dropt at present, since the proprietors of the mines have found that their, copper ores, containing silver, can be with less expence carried to and parted at *Thajola* in *Lower-Hungary*. The whole annual produce of *Oravitza* is about one hundred and fifty of copper.

*Saska* has four furnaces, called *Charles, Joseph, Maximilian,* and *Radimer-Hutte. Moldova* has but one. The process in both places is the same as at *Oravitza*; but the vicinity of large forests makes it less expensive; and the great plenty of copper mulm found at *Saska* makes it there very easy. The *Saska* and *Moldova* smelters boast of their smelting the ore with an increase of copper, its common assays giving only three or four pounds. This for a long while seemed a riddle to me; but I fancy with some reason,

son, that this additional produce is owing to their additional pyrites.

I have told you already that *Moldova* produces the toughest and most malleable copper. This seems owing rather to the sulphurous nature of the ore, than to any particular advantage in the smelting.

*Moldova* gives per year about 50 tons, and *Saska* about 150 or 200 tons of copper.

The *Dognazka* ores being greatly sulphurous, their smelting and refining is less expensive than at *Oravitza*, though the process be entirely the same. There are three smelting places, with ten furnaces. The ores, smelted promiscuously, and in common as at *Oravitza*, yield every year about 200 tons of copper. These ores, containing less than nine ounces of silver, cannot be parted in our inland furnaces; * for this reason the proprietors have tried many experiments of other more successful proceedings. Even at present the surveyor, Mr. *Fluk*, has proposed a plan to precipitate the silver in the smelted

---

* This does not give any credit to the skill of the *Hungarian* smelters. The copper and lead ores of the famous *Ramelsberg* near *Goslar* in the *Lower Hartz*, do contain but about one ounce of silver; and are besides greatly refractory. However, they are with great advantage parted in the furnaces belonging to these mines.

copper, and by this means to save it. It would prove a great advantage to the proprietors. Mr. *Delius* has likewise proposed and advised some alterations in the smelting, to deprive the *Dognazka* copper of the brittleness, which the coppersmiths complain of.

Some days ago Baron *Hegengarthen* received an account of the goldwashings in the *Almash*, and orders of the court to examine them. Counsellor *Koczian* is author of this account, and I will send it to you as soon I shall hear of the result of their examination

The evening before I set out from *Oravitza* we had a terrible thunder storm. I happened to stand at the door, and to see under a violent lightning a flame rising behind an opposite house, which keeping itself some time at its top rushed at last down on its foreside, and then returned to the place whence it first arose. This phænomenon was repeated several times. We examined the place whence this electrical evaporation came from, and found that pyriticous fissures lay hid under the vegetable mould.

My journey from *Dognazka* to *Bogsham*, and thence to *Lugos*, is one of the most singular I ever made in my life. The danger of the roads caused Baron *Hegengarthen*, whose humanity you know, and whose kindness I never can praise too much, to send orders for my safety wherever I had to pass. Accordingly I found in every village forty or fifty *Wallachians*, armed with firelocks, who under the conduct of their chiefs escorted me to the next, and in rough or stony roads did bear my coach rather than support it with their shoulders. The same day there was a general chace in the country to surround the forests and to search after the robbers. This is done once every year, but commonly without any success, since the requisite orders cannot be kept a secret from the robbers, who for that reason stay quietly at home that day, or even dare, in compliance to the orders, to follow the general chace. *Bogsham*, where I dined, is but four hours journey from *Dognazka*. This place is situated in a fine valley, surrounded with clay slate and limestone hills, superincumbent on our metallic rock (*Saxum metalliferum.*) The river *Berfova* runs through it, but the adjacent fens and swamps make its situation very unwholesome. As *Servia* continued under the imperial jurisdiction, this place had many fine buildings and iron hammers; but now the iron trade is stopt. Nevertheless there are still some iron hammers, and furnaces; and vast quantities of bullets and shells are cast here for the imperial artillery. The iron ore comes from *Dognazka*. It is either red ore, *ochra ferri indurata rubra*; or black, *ferrum refractorium tritura atra textura chalybea*; and

and gives a good iron. Near *Bogſham* is a calcareous hill at a place called *Valga baja,* which contains immenſe quantities of broken oyſter ſhells and mytulites. From *Bogſham* to *Lugos* is a continual ridge of granite hills, below ſhivery micaceous clay. From *Lugos* they run to the eaſt towards the high mountains, which ſeparate *Tranſylvania* from the *Turkiſh* dominions. The calcareous hills about *Lugos* produce good wine; and I taſte it at preſent to your health

A P P E N-

# APPENDIX I.

FROM

MR. CHRISTOPH. TRAVGOTTS DELIUS,

Affeffor in the Direction of the Mines in the Bannat:

A Propofal to foften the Copper, prefented to the Imperial and Royal Chamber Court at Vienna, *July* 16, 1768.

THE copper ores, whether mineralized with fulphur or arfenic, or with both together, contain, befides the copper and the unmetallic earth, a part of iron; and they are diftinguifhed only by its greater or fmaller proportion. The yellow pyriticous ores for example, the rainbow coloured, the fallow, the copper and the glafs ores, and in general thofe that are remarkably mineralized with fulphur, contain more iron than the green and blue ores; but the copper ochres and liver ores contain more of it than any other fpecies. This iron mixture, if in the fmelting and refining it be not entirely removed, is the proper and real caufe of the brittlenefs of copper; and though it be likewife produced by arfenic, this however happens

pens only if this half metal be united with iron, since by itself, and unconnected with iron, it is too volatile to resist the intense and repeated heat of the copper preparation, if the roastings and smeltings be properly directed. Therefore the great principle to get a fine, malleable and soft copper consists in its careful separation from the iron; and the usual practices are chiefly aimed at it, though by the following reasons they fall more or less short of their intention.

It is a known fact, that nothing destroys iron so fast and efficaciously as sulphur. Though commonly the copper ores contain a good deal of sulphur in their mixture, it is insufficient by itself to destroy the admixed iron. Therefore a certain quantity of sulphur pyrites is added to every first smelting of copper ore, to get its sulphur mixed with the iron, to have it by this sulphur in the ensuing roasting of the lech or raw stone calcined as in a cementation; and lastly, to have the remaining sulphur and iron scorified either in the smelting of the black copper, or in its final refining. This method of smelting, invented by our ancestors, is in the main so well adapted to nature, that with all our refinements we are at loss to invent a similar or a better. It would certainly and perfectly answer all its ends, if there was not a circumstance which causes difficulties, and is proved by metallurgic chemistry; and that is, that all sulphur-pyrites contain a good deal of iron.

iron. Accordingly what is procured by one side is in a certain respect lost again on another. Its sulphur may very well be supposed to destroy the iron particles of the copper ores; but as it contains a good deal of iron in its own mixture, its sulphur is insufficient entirely to destroy the iron in the compound mass, a part of which unites unaffected by it with the *lech* or *rawstone*. In the ensuing roasting the sulphur, which remained in the rawstone, together with a part of the iron, is destroyed by cementation; the former evaporating, and the latter changing into dross, which in the following black copper smelting is taken off with the flags. On this account a part of iron remains in the black copper, which in the last refining cannot be entirely destroyed, as then scarce any sulphur is remaining. Hence it comes to pass, that provided there be no want of good intelligent smelters, those places produce the best copper, where the ores contain but a small quantity of arsenic and iron, and where they have plenty of good sulphurous pyrites. If any pyrites was to be found entirely destitute of iron, it would undoubtedly produce the most excellent and ductile copper. But as that is not to be expected, and nature has not favoured our wishes, we are to look about us for other means to soften the copper; for which reason, and the encouragements granted by her Imperial and Royal Apostolic Majesty,

F  I ven-

I venture to propose some of my ideas, established on many assays and experiments. However, before I come to my manner of refining the copper, I shall lay down some rules, which are highly subservient to the purpose, and should never be neglected by those who are at the head of any great metallurgical works.

*Primo.* As I have shown already that pyrites, containing much sulphur and a smaller quantity of iron, proves an advantage in the first or raw smelting, care is to be taken that such pyrites, and not indiscriminately any other, be chosen for the raw-work. No pyrites is to be made use of which is both arsenical and sulphurous, since arsenic unites with the iron, and causes a great brittleness. For that reason the furnace inspectors ought to examine their pyrites and their constituent parts, which commonly is neglected; since for the most part being unskilled in the operations of metallurgic chemistry, they are unable to make such analyses. However they assay them with proper fluors for lech and stone. But the produced lech or stone-grain being a compound of sulphur and iron, it is impossible to know by this useless assay how much sulphur and how much iron is contained in the pyrites, and whether any arsenic is united to them. Sublimations in closed vessels are preferable in every respect; and smelters and assayers ought to chuse this method, since in the

assays

assays of every sort of ores and minerals it certainly is more instructive and precise than the common empirical assaying. If by these means good pyrites be procured, proper care ought to be taken of the proportion, in which it is to be added to the first smelting. This is to be determined by the quality of the ore; if irony and refractory the addition of pyrites ought to be in a larger quantity.

But many faults are committed on this account; since many furnace inspectors, by a misapplied œconomy and to save some pyrites, or to have the rawstone rich, and to save the trouble and expence of roasting, grudge the addition of pyrites, and by that spoil the nature of the copper. There is never any real advantage in the riches of rawstone, since it impoverishes the mass of sulphur, which in the ensuing roasting is insufficient to destroy the greater proportion of iron. A good lech or rawstone should, to produce ductile or good copper, never contain above seventeen or eighteen pounds of copper in an hundred weight.

*Secundo.* If a copper mine produce pyritical and sulphurous copper ore, which, without any addition of other unsulphurous ores, are smelted with sulphur-pyrites as usual at *Smólniz*, it is greatly productive of ductile copper to have both ores and pyrites gently roasted before the first smelting. The reason is as follows: Sulphur does

does not destroy the iron, by its combustible matter; since consisting of vitriolic acid and phlogiston, the latter evaporates by a gentle roasting. But the vitriolic acid penetrates into the iron, dissolves it into a crocus and destroys it, which causes it in the ensuing smelting to go easily off with the slags, and to leave the copper regulus or the rawstone in a more depurated state. Though this be an undoubted principle of rational metallurgy, I fear it will be objected to by some smelters, who know only ancient practices. Would they please fairly to try some experiments they might be convinced of its utility. However, it is to be observed *first*, that this rule is not general, because, if refractory or unsulphurous ores are to be smelted together with the more sulphurous ones, the roasting is impracticable, since the vitriolic acid and the sulphur of the latter, requisite to the fusion of the former, might be inconsiderately destroyed by it. *Secondly*, I have said for very good reasons, that the roasting ought to be gentle. A violent roasting might smelt the ores and unite the sulphur to the iron and to the copper; and it is a known fact, that sulphur, destroying iron in a gentle cementation, by a strong fire, is brought into fusion with it. Besides an intenser roasting would destroy too much sulphur, which, after the raw smelting, is likewise intended to scorify the deaf substances of the ores.

*Thirdly*.

*Thirdly.* On these principles there might be laid down many improvements of the common roasting of lech or rawstone. Any roasting intended to be useful ought to be gentle. The object of roasting rawstone is partly the evaporating of the combustile sulphur and of the volatile arsenic, so as to bring the mass of ore closer together, and to facilitate its subsequent fusion into black copper; and partly the producing of the sulphurous acid and its calcining and destroying the iron of the rawstone. This double object is better obtained by gentle than by violent fire. In a strong fire the ores coagulate and smelt together; the arsenic uniting with iron is fixed, and both make with the fluid sulphur and copper a compound mass, which is hard to part again. To be convinced of it, melt iron, sulphur and arsenic into a regulus, pulverize it and expose the powder to a gentle fire; you will find these minerals evaporated in a short time, and the destroyed iron remaining in the crucible. But put the same in a smelting fire, you will find the whole for many hours in fusion, without any remakable decay or destruction.

To obtain the ends of gentle roasting the following rules are to be observed:

1. The roast ought never to be in open air, but to be included by walls and to be sheltered, to shut out the irregular blowing of the wind and rain and snow,

snow, as either producing by intense fire a partial coagulating and smelting of the mass, or by interrupted fire an unequal and imperfect roasting, which in either of these cases puts a stop to the intended evaporation of noxious minerals, and the destruction of the iron.

2. The masses ought not to be too large. The common practice at present is to roast one hundred and eighty or about two hundred hundred weight of lech or rawstone together; nay, some furnace inspectors, to save a trifle of charcoals and wood, go still above that quantity; but this grudging œconomy spoils the copper, since the greater the heap the greater the fire, which is produced by the greater quantity of sulphur, and causes a smelting and coagulating in the mass, whose obnoxious effects have been touched upon before. More and smaller masses, each of one hundred, or at most one hundred and twenty hundred weight, might be preferable; so it will prove equally conducive to spread the ore in thin layers over the alternate wood, and only to employ in the first fire as much fuel as might be requisite to a gentle roasting, which in the following fires is constantly to be kept gentle and equal.

3. In some cases the double rawsmelting is to be considered as a great advantage to produce ductile copper. Innoxious to any ore, it might be superfluous with some sorts; but it is highly useful

useful and necessary for those that are arsenical and irony, or refractory on account of brittle and deaf minerals or half metals. It is impossible by the first smelting to remove the greater part of these noxious minerals, and it is equally difficult in the subsequent roasting and black smelting to deliver the metal from their influence. Therefore two or three gentle roastings of the lech or stone of the first rawsmelting, and its subsequent second smelting with a proportionable quantity of pyrites into a *double lech*, are to be greatly commended, since the sulphur of the pyrites unites in the second smelting with the destroyed minerals, calcined by the preceding roastings, and carries them off in flags. This *double lech* is then to be brought to a regular roasting, and afterwards smelted into black copper. But in this process the rule of an equal gentle roasting is more carefully to be observed than in any other; else the whole will run together, and the wild and deaf minerals so closely unite to the copper, that it will prove almost impossible to separate them again.

As it is an acknowledged rule, that only *good black copper* produces *a good fine copper*, I do not dwell any longer in recommending the preceding general rules, and the careful roasting, which is too much neglected. By what I have said it clearly appears, that the fine copper will be brittle and irony, in the same proportion as the previous works have been neglected. However, whatever care be taken there will still remain some iron, even in the refined copper. The reason is this: A hundred weight of good black copper contains commonly about ninety pounds of fine copper, and about ten pounds of iron, or sometimes arsenic united with iron. The trifling quantity of sulphur still existing in it destroys during the refining a part of this iron, and even the fire scorifies a part of it; but as generally the iron in the black copper is equal in weight if not superior to its sulphur; this little sulphur is certainly unfit to scorify it entirely.

Therefore I have supposed, that during the refining something might with great advantage be added to the copper, to purify and to soften it entirely.

Two mineral substances destroy and scorify iron, litharge and sulphur.

The former is unfit for the refining on the *hearth* (*Gar-kerd*) since the glowing coals which cover the copper might reduce it into lead, and this uniting with the copper cause it to be leadish; but in a *parting furnace*, where the copper is kept in fusion by the flame, it might do, since wood flames do not reduce it, facilitating rather the calcination of lead, and leaving it floating as litharge on the surface of the copper. Being under these

these circumstances in a continual motion on the surface of the boiling fluid copper, it will attract the remaining iron particles, scorify with them, and leave the copper in its highest purity. The quantity of litharge is to be determined by the lesser or greater quantity of black copper. Six or eight pounds per hundred weight might do.

Concerning the refining *on the hearth*, the best effect is to be expected from common officinal sulphur. The iron has no greater enemy, and it does not affect any other metal as long as it has iron to work upon and to adhere to. Accordingly, as soon as the black copper is brought to smelting fusion and boiling sulphur in pieces ought to be put on its surface, and covered with coals to concentrate the boiling. This may be repeated two or three times; but the compound sulphur and iron flags are carefully to be taken away, and no more sulphur to be added, than what is requisite to destroy the iron; else the superfluous sulphur uniting with the copper causes unnecessary expences and lengthens the refining. For this reason the black copper is previously to be examined. In proportion to the iron contained in the black copper, three, four or five pounds of sulphur will be sufficient to its destruction, and to the copper's highest refining and softening.

This sulphur-refining will equally do in the parting furnace.

These

These proposals are, for what I know, entirely new and never practised before; however, they are so adequate to the natural rules of metallurgy and so cheap, that I dare hope they will not only stand the trial of impartial intelligent smelters, but prove likewise highly conducive to the production of fine, soft and malleable copper.

## POSTSCRIPT.

It has been demonstrated before, that the copper will be less irony in the same proportion as the pyrites added in the first rawsmelting are destitute of iron. But as the pyrites are commonly to be used as they may be had, and generally any sort of them contains a good deal of iron, the following process is recommended as an improvement of the first rawsmelting.

At present the pyrites is added to the rawsmelting of the ores in its natural form; but it would be better to smelt it separately, and to bring it previously into a sulphurous regulus, since the greater part of its iron will be by such a smelting brought off in dross, and its lech or regulus consist for the greater part of a pure sulphurous mass. This sulphurous regulus I propose to mix with the copper ore in the first rawsmelting; and half the quantity of the usual natural pyrites will

be

be sufficient to produce a less irony copper regulus, which according to my rules, and properly roasted, will undoubtedly give a very malleable copper.

The objection, that this separate and preparatory smelting of the pyrites will cause a considerable expence is of no great weight with me; since the pyrite regulus being a milder and easier fluor than the unprepared raw pyrites, the fire and time requisite to prepare the pyrite regulus will for a great part be saved again in the easier and shorter rawsmelting.

APPEN-

## APPENDIX II.

Obfervations on the Goldwafhings in the Bannat,

### BY COUNSELLOR KOCZIAN,

With the refult of the enquiries made after them

### BY MR. DEMBSHER.

AMONG the feveral natural advantages of the *Temefwar Bannat* fome of its rivers are known to yield gold-duft. I could not neglect this object when lately I travelled in thefe parts.

The goldwafhing in the Bannat is properly the bufinefs of the gipfies *(Zigeuner)* and left as it were to this poor people as an exclufive trade. This laid me under the neceffity to apply to them for inftruction.

The river *Nera* in *Almafh* carries gold-duft, and feemed to me the fitteft for my purpofe; accordingly I caufed fome gipfies, reputed to be very fkilful, to make a wafhing near a village called *Bofhowiz*; and I faw with pleafure, that with much dexterity in a few minutes time, they cleared in the trough the value of fome grofhes of of gold; they showed me likewise among their gold-dust some pieces of a remakable bigness.

After having sufficiently observed and examined their simple maniputations, which I shall speak of more in the sequel, I wanted to know the origin of this river gold.

A particular circumstance favoured my curiosity. I saw that the gipsies washed it from the sands not only taken in the river, but likewise from its borders, nay even from some pits in the adjacent ground. These pits are commonly four foot and more deep, and yield richer sands than the river itself. They told me likewise, that the river sand grows richer in the same proportion as the waters are high; and that it is poorer in dry weather. Such it was in 1769, and consequently they were forced to open the goldsand pits in the adjacent grounds.

I examined these pits and the country around the *Nera*, which has been delineated in the annexed plan.

The strata on its borders are as follows: The first is common vegetable mould, nearly of one foot thickness; the second loam, two feet; the third pebbles and calcareous earth, hard to be dug with pickaxes, one foot and a half; the fourth or the goldsand bed is three feet, consisting of a mixture of pebbles, rockstones and fine irony sand. This last stratum is the same which the

gipsies

gipsies, at thirty fathoms distance from the river, dig out for washing. According to what I could see in the pits this bed has a slate bottom; and somewhat lower down the river a large coal-stratum bassets out. I might therefore say with some probability, that after the slate follows clay, then marle, and afterwards the coal bed.

From all this follows, 1. that the gold-dust is not generated by the water, but brought in the river beds by accident, because in the former case it ought to be found in constant and equal quantities, whether the water be high or low. 2. It ought to be had from such beds which may be easily dissolved by water; accordingly it is not owing to sound veins, since rain water and torrents cannot possibly be supposed, in their short and intermittent flowing, to carry off even that part of gold which they commonly leave behind; and besides they would have long since discovered the veins in the many countries where gold is washing, and where no such veins have been found out. 3. Therefore the gold-dust is probably owing only to clay and earth beds, dissolved and carried off by water.

The bed which the gipsies dig out is of such a dissoluble nature; it is gently dipping, and by what I could see ascending or rising from the west to the east. Being in this hanging or gently dipping situation, it may possibly be laid bare in several

veral parts of the river borders, and washed off by high water, which very well explains the greater success of the washings after heavy rains.

On my further journey in the Bannat I observed many marks of old washings, probably left by the *Romans*. They pursued likewise the gold impregnated beds, which in many places must be six fathoms above the river borders. Near *Werſherova, Polvaſhniza, Purlava, Tumul* in the *Karanſebez*, and in the valley *Walle-mare*, towards the limits of *Tranſſylvania*, from *Obava-Piſtra* till *Marga*, it is plain that they dug for gold in such elevations, which never could be reached by the river water. In *Tranſſylvania*, near *Olah Pian*, at the foot of the *Rudel* mountain, many old gold-pits are found in a dry country, which is entirely deſtitute of brooks and rivulets.

This clearly shows that the gold impregnated beds are not to be considered as river sediments deposited on the borders. § If they were succeſſively accumulated and washed off from the adjacent hills, there is no reason why the gold should be

---

§ This may be and certainly ought to be granted, in reſpect to the preſent viſible brooks and rivulets; but may be with equal juſtice denied in reſpect to thoſe of former times, since the ſurface of the earth has undergone ſo many ſucceſſive revolutions, and ancient ſeas, lakes and rivers are every where traced in the preſent continent by their former effects. (Tranſl.)

only

only contained in a single bed, and never to be found in the upper vegetable mould?

The solid compact stratum of dragged pebbles and rocks, which is superincumbent on the gold bed, is a further argument for its not having been produced successively; since no reason appears why the gold-dust might have been carried and deposited under this compact stratum. ‖

Therefore we have the greatest reason to believe, that the gold impregnated bed is owing to the deluge, and that accordingly it is wide, stretching through considerable tracts of land. * In this supposition remains but a single question: whether this bed be throughout impregnated with gold? Though this might be affirmed for very good reasons, I will however, to corroborate it, take notice that the *Romans*, beginning their washings near the river, continued them a hundred fathoms length in the adjacent lands, and as long as they could reach and easily lay open the gold

---

‖ These two inductions are extremely precarious. Might not successive revolutions, whatever they were, produce different strata of a different nature? (Transl.)

\* The latter is fact; and the diluvian supposition a bad consequence drawn from precarious inductions as well as from too narrow principles. (Transl.)

bed

bed without driving galleries, which seem to have been unknown to them. ‡

By similar circumstances the gipsies have no chance to make any greater progress, being confined to the gold which is carried by the rivers, or contained in the less incumbered and buried gold-bed near their borders.

However, the object being of importance, and deserving nearer examination, I should advise to drive a gallery in the gold impregnated bed, and to examine how far it runs into the field, and whether it constantly keeps the gold impregnated quality? If it should be found to extend a considerable way in the mountains, and to continue gold impregnated, it would be worth while to have large washings on regular hearths.

The present manipulation of the gipsies is as follows: They use a board of lime tree one fathom length, and one inch and a half thick. At the upper end is a small trough, and across the board are ten or twelve small cuts or furrows. This board they raise at one end, under an angle of nearly forty-five degrees. The sand is put in the trough at the upper end, and

‡ Mr. Koczian is very unhappy in suppositions. Had he never heard of the *cuniculi* of the ancients? What are they but galleries?

thence by plenty of water washed down the sloping of the board. This causes the lighter sands to be washed off, and the heavier ones to remain in the furrows and on the surface of the board, whence they are scraped or brushed off, to be separated from the gold by the operation of the common trough. Their whole proceeding is so extremely careless, that a good deal of gold is lost by it; and what is still more to be pitied they get but the pure gold-dust, that which is still sticking to the sands and stones being thrown away, as I am convinced by the microscope, nay even by simple ocular inspection.

This circumstance deserves nearer examination: whether these sands and stones be rich enough to bear the expence of pounding? A small trial might be sufficient. If they should appear to bear it, regular mining and pounding would be adviseable.

I cannot conclude without adding a particular observation, which I had an opportunity to make near the many old and new gold-washing places in the Bannat, and which I consider worth attention. I found that the higher promontories on the gold impregnated rivers do not consist of solid rocks, but of soft earth-beds, which give good indications of coals and alum ores. Near *Boshowiz* on the *Nera*, which is known by its gold-washings

washings, I saw a large coal bed baffetting out at a small distance from the gold sand bed; and almost in every place, which I have spoken of before, the exterior appearance of the ground countenances the conclusion, that coal beds are below the gold impregnated stratum, and that these in a certain respect are to be considered as its sole. They had perhaps some share even in the generation of the gold; at least, they have a great relation with gold, since it is not impossible to extract from them a hepar sulphuris, which is the strongest dissolvent of gold.

However this be, it is fact that coals are to be found in every gold washing-ground. The *Danube* and *Ens* may stand an evidence, since on the borders of the *Danube* from *Vienna* to *Passau*, nay still higher up, coal beds offer every where. Therefore I do not doubt, that the same gold impregnated stratum may be traced out on the borders of these rivers, and regularly worked to advantage.

According to these observations and accounts of Counsellor *Koczian*, and the orders which I had received, I proceeded to my enquiries in the following manner:

As soon as I arrived at *Boshowiz* I enquired after the place, in which Baron *Koczian* had made his observations; and I found it as laid down in

plate

plate   to which this explanation is belonging.

A. *Menish*; a brook.

B. *Boshowiz*; a village.

C. The ground, where Baron *Koczian* caused his washings. It rises gently towards the mountains; and

D. a gallery of one fathom length was drove into it.

E. The coals in different beds.

F. Mouth of the *Menish*, where it runs into the *Nera*.

Near K and L the ground is very flat; and consists, as appears in some higher borders, of different stratified earth and stone beds.

After a general survey I caused the works at D to be cleared, to get acquainted with Baron *Koczian*'s gold impregnated stratum and its foundation. I found it agreeing with his description, and consisting of a mixture of brown loam, pebbles, rocks, mica, garnets and iron sand. But the under bed was no slate, consisting rather of a brown sandstone, extremely mouldering and friable in the pit, and hardening in the open air.

Acquainted with the object of my enquiries, I advanced 27 fathom nearer to the mountains, and orderd a shaft to be sunk in G. In a depth of one fathom and a half I reached the gold impregnated stratum. A washing convinced me of its
containing

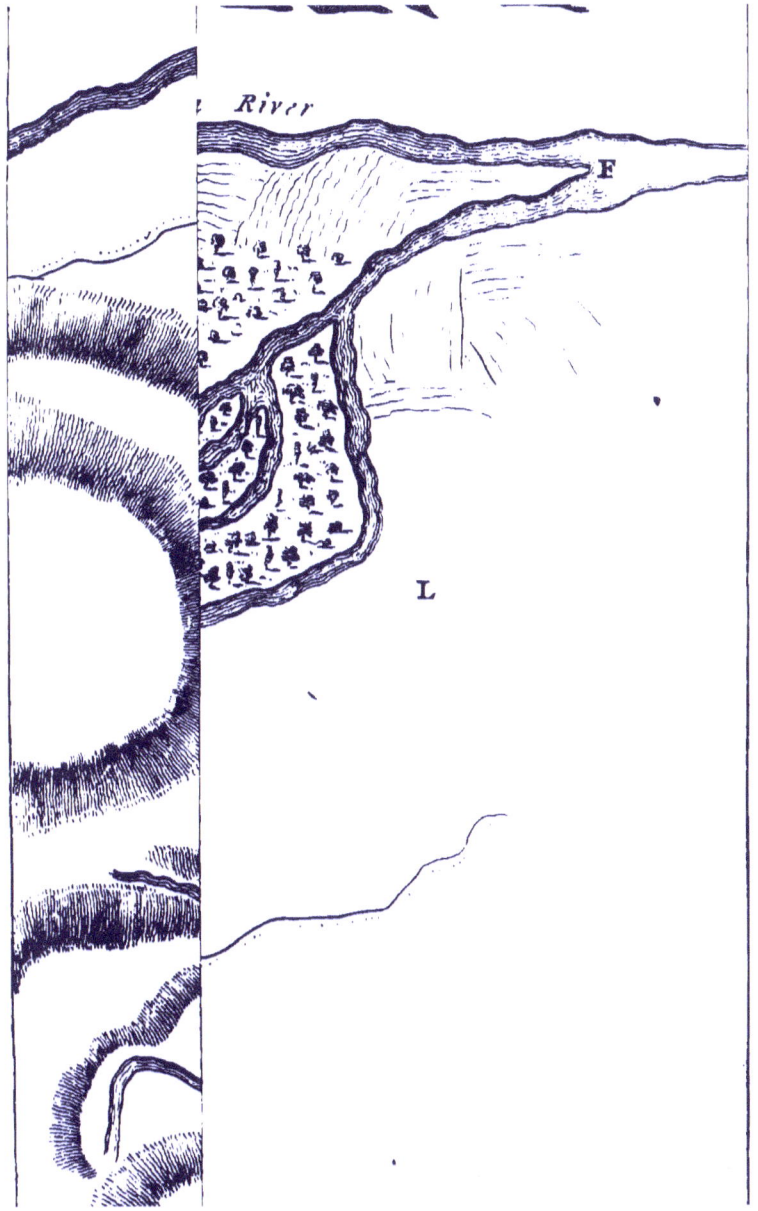

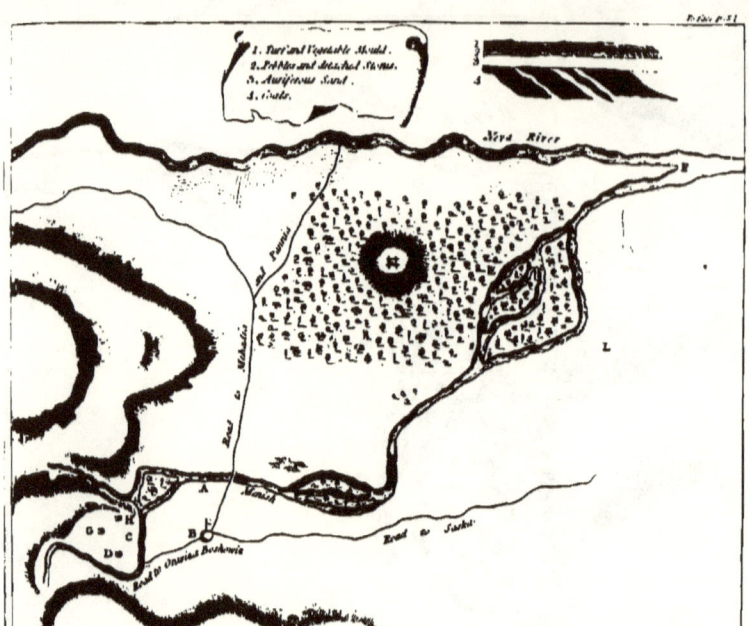

containing gold; and I ordered thirty carts load to be laid afide for a great wafhing on the hearth. I advanced then ftill twenty eight fathom more, almoft to the foot of the mountains; and to be convinced of the extenfivenefs of this ftratum, I caufed a fhaft to be funk in it, twenty-two fathom to the fouth. Here I found the beds entirely different. The grey loam immediately under the turf was very tough, and a fathom thick. Then followed brown loam five foot, afterwards the compact pebble bed four $\frac{1}{2}$ foot, and then the gold impregnated ftratum. After a little trial in the trough, I ordered as much to be laid afide as might be full fufficient for a wafhing at large. Then I proceeded to thefe wafhings on a hearth, exactly conftructed as thofe at *Shemniz*. The refult was as follows:

*Firft proof.* Thirty carts from the fhaft G yielded two grains of gold.

*Second proof.* Thirty carts from the fhaft H yielded fcarce half a grain of gold.

The greateft care was taken, and I was too well convinced, that fuch profits could not bear the expences of mining, which, as here to be undertaken immediately under the turf, would require a good deal of timber, of which the whole adjacent country is entirely deftitute.

Therefore I dropt my works in this place, and examined the bed which is gently rifing on the coal ftratum along the *Menifh*, and in feveral places is three fathom above the water level. I conftantly found fome gold flakes in the trough; but in fo fmall quantities, that I faw no encouragement for a wafhing hearth, and I gave up every hope of mining. The common gold-wafhers having for the moft part retired to *Banya*, *Ruderia* and *Telpofhiz*; I followed them to thefe places to examine the ground and earth which they were wafhing there. At *Telpofhiz* I found it as in the before defcribed places; but at *Banya* and *Ruderia* I faw the gipfies feeking for gold in the gutters and furrows of the mountain-brooks.

So much for the hiftory of my enquiries. I fhall add fome obfervations produced by them, and explanatory to thofe of *Counfellor Koczian*.

1. As foon as a fhaft or drift reaches the gold impregnated bed you reach water. This is fo general that I have found it fo in the fhaft H and E, three fathoms above the water level of the *Menifh*.

2. The gold wafhed hereabout is entirely native, free from any matrix. It appears in fine duft. Tho' the exterior appearance of this mixed ftratum convinced me, that it is not owing to gold veins and fiffures, and that accordingly its rocks and ftones cannot poffibly contain any gold, I caufed however for my fuller conviction the wafhed fands and

and stones to be stampt and washed again. No gold appeared. I then had them roasted by fire, but without any better success.

3. The deeper this bed under ground the richer. It grows poorer in the same proportion in which it ascends to the mountains; which in a certain manner is to be explained by the first observation.

4. This gold impregnated bed yields every where a pure black splendent sand, which might be called perhaps native iron, since it is drawn by the loadstone.* In the gutters and furrows at *Banya* and *Ruderia* but an insignificant quantity of this iron sand is found; they give a greater quantity of pyriticous sand, which, together with the ores and gang-rocks, is a strong presumption for copper mines. The fine situation of these mountains, water, wood and timber being plenty, might give a zealous miner a mind to explore them.

It is very difficult to determine the origin of gold-dust contained in this stratum. Though the discovered beds, the extensive coal stratum, and now and then some petrifactions, be strong evi-

---

* The translator has verified this observation in the gold-washings on the *Eder* in *Hesse*. He might perhaps give a natural reason for this iron sand, which constantly is found concomitant with the gold-dust.

dences of great inundations; there appears no sufficient reason why the gold dust should be only mixed in the substance of a single bed? As I am no friend of conjectures, I leave the explication of this phænomenon to men of greater genius, enabled by their extensive knowledge and experience in Natural Philosophy, to make discoveries beyond my capacity.

I am to answer here to an objection which seems to be a just one, and is, that notwithstanding the poorness of these sands, the value of some thousand florins of gold is produced every year. Though this be fact, and a considerable sum in itself, it is a very inconsiderable one in respect of the great number of people employed in the washings. There were for example in the year 1770, in the neighbourhood of *Uy Palanka*, *Orsowa* and *Caransebez*, above 80 families of goldwashers, men, women and children, employed in that business; and nevertheless they have not made good above the value of six or seven hundred ducats. Hence it appears to me that these washings are no objects for miners, and less so for *Germans*. The gipsies go half naked; whole families live at the daily expence of a groat, nay cheaper. Satisfied with this petty allowance, and unconcerned at their nudity, they wash gold in summer time, and during the winter they cut wooden troughs, spoons

spoons and the like, which they ramble about with selling and begging. A miner would scorn such a life; and if you would keep them better, whence is to arise the profit of the sovereign, which in the common way of the gipsies is something, though it be inconsiderable?

Concerning their manipulation it seems to be at first sight very rude and bungling; but it is very just in itself. Practice has given them an experience, destitute of which one might consider their process as very deficient. I am convinced of it by its following examination. When they had finished a washing of fifteen or twenty troughs of sand on their usual board, which is seven foot in length, rifled with fifty or sixty transverse furrows, and erected under an angle of eighteen or twenty degrees, I caused the sands, which remained in the furrows, to be divided into three parts. The greater part of gold stuck constantly in the ten or fifteen uppermost furrows; in the ensuing division I scarce met with the eighth part of the former; and in the last fifteen or twenty furrows scarce two or three flakes of gold were to be found. I have likewise carefully examined the sands, which they had washed already, and it was but very seldom that any mark of gold was left in them.

Such

Such was the result of my first journey, which, contrary to my expectation, produced a second. Unluckily I had not clearly enough explained my third observation; and Baron *Hegengarthen* was hence inclined to believe, that the gold impregnated bed might prove richer in a greater depth. Therefore he proposed to sink in K a shaft of nine or ten fathoms depth, till a solid sound and barren rock might be reached; and I was accordingly ordered to return with two able workmen in the *Almash*, and to try the experiment.

It was *June* 13th in 1771 when I did so. The point K is about three feet above the water level of the *Menish*, which having in E torn off a steep part of its borders, and by that accident laid open the coals and the other strata, I could with certainty foretel that the sole of the gold impregnated sands would prove to be either the coal or the alternating marle beds. With the same probability I might have foretold the bigness of these strata.

After $\frac{1}{4}$ fathom of vegetable mould I reached the first sand and pebble bed and water, which increased, arising from the sole, in the same proportion as we sunk deeper. Two troughs of rubbish were had with five or six troughs of water. This circumstance and the constant rains filled my unsheltered works with so much water, that by twelve hours labour I could not get but three foot depth. After one $\frac{1}{4}$ fathom I reached the gold

gold impregnated bed, which I very often examined on the trough, but constantly found so poor that a large mine trough contained often but a single flake, and often nothing at all. At last we fell in with the coals, and having sunk my work $\frac{1}{2}$ fathom in them to no purpose, and found that the beds were the same as those by accident laid open in E. I gave up my works in this place, and examined rather these discovered strata. The result was the same as that of my former enquiries.

My many small washings did not encourage me to further experiments; and thus I dropt my repeated enquiries.

To prevent all further doubts I add an observation, which proves with nearly mathematical evidence that this side the *Menish* only coals and marle, and on the other side towards the mountains only slate and sand stone will be found by digging deeper.

The gold impregnated stratum is constantly parallel to the turf and vegetable mould. It does not answer at all to the dipping of the lower stone beds. This appears clearly in E, where marle and coals, alternating with regularity, are never parted by earth beds but covered by the gold impregnated stratum, which is parallel to the turf. The section plate explains it to the eyes.

Here

Here the question arises again: why this gold was produced in this stratum? I freely confess that I am at a loss to answer it; but I cannot entirely abstain from proposing some conjectures, which perhaps may assist others to discover the secret.

The hypothesis of those, who explain the origin of coals by forests, buried and swallowed up by earthquakes, gains some credit by the exterior appearance of the *Almash*, whose soil is every where intersected by hills and rivers, and is entirely destitute of wood. But the remarkable regularity in the alternating and parallel coal and marle beds does not agree with the idea of such dismal and violent destructions. The structure of the uppermost loose soil in these parts is more easily to be explained by inundations. It is a known fact that gold and iron are generated and produced in many flat countries, and that these metals are dug in many places besides the common veins and fissures. The Dutch sea sand and the iron, which in different places is found concomitant with the gold dust, are strong evidences.§ Supposing this native gold to have been contained in the uppermost and looser beds of the

§ Of what? That the gold or iron is generated in these beds? No; that they are commonly found together, and that having perhaps had a common origin, they have been washed and deposited there by the same revolution or natural cause, whatever this be.) (Transl.

hills,

hills, it could be thence carried by inundations into lower grounds, and as inundations do not retreat but succeffively, gold, iron, garnets and sherl ought of courfe, and according to their fpecific gravity, to have taken the loweft place, and to be depofited in thofe flats, which as the loweft have been interfected by the rivers.

This opinion agrees exactly with experience, and explains the reafon why the gipfies get greater quantities of gold in high water that in dry weather. By heavy rains the borders of the rivers and brooks are worn away; nay, under fuch circumftances, the rivulets take often a quite different courfe. This faciliates the manipulation of the gipfies, becaufe the water diffolving and carrying away the argillaceous particles, and leaving only the heavier fands and the gold-duft behind, they get by a fingle trough of fand as much gold as they might have wafhed from two or three troughs of the undiffolved natural fands.

But is this an explication of the orgin of this gold duft? Certainly no. However, I have done what moft naturalifts do, I have advanced my opinion.

FRANCIS DEMBSHER.

LETTER

## LETTER XI.

*Nagyag, July* 12, 1770.

THE plains, which I reached near *Lugos*, continued half ways to *Dobra*, where I found an afcending ground of argillaceous flate. Beyond *Dobra* I met again with our *Saxum metalliferum*. It continued to *Deva*. The roads are impaffable. Between tremendous precipices and the river *Maros* on one fide, and fteep fhaggy rocks on the other, I was dragged along, by eight oxen, added to the four horfes of my carriage. I arrived late in the night at *Deva*; but the fafety of the roads made amends for their roughnefs. As foon as I came to the limits of *Tranffylvania*, between *Dobra* and *Deva*, the two huzzars, which I had taken at *Lugos*, left me. The *Tranffylvania Wallachians*, more humanized than thofe in the Bannat, and the national frontier troops, together with the feverity of government againft the robbers, contribute greatly to the fafety of the country. It is but of late that three robbers have been at *Deva* impaled alive, for having committed fome murders in the valley of *Hazeg*. This cruel and

and almost inhuman punishment, tho' used in *Slavonia* and the *Bannat*, has made such an impression on the inhabitants, that you may travel all the night with safety. The day after my arrival I visited the copper-mines, which some years ago have been opened in a mountain to the west, three quarters of an hour's journey from *Deva*. The mountain consists, at the foot, of a micaceous slate, covered with indurated marle rocks gently rising. In these rocks are the fissures, which compose the copper stockwork at *Deva*. I desire you would understand the word stockwork in the same sense in which I explained it to you in my letter from *Dognazka*. However, there is a great difference between the stockworks of these two places. That at *Dognazka* consists of large and very rich veins, converging and uniting in the same point: here it consists of some fissures, uniting in a middle of ten fathom diameter, and mixed with dead rocks. They have pursued the run of the fissures, and sunk in it a shaft of some fathoms depth, but without any success either in the drift or depth. The vein *(gang-art)* is grey and loose clay, now and then sprinkled with quartz and spath, and containing various coloured and yellow copper-pyrites, which, if of the richest sort, contain seventeen pounds of copper per hundred weight. A hundred weight of this copper contains one dram and two denarii of silver;

and

and a mark of this silver two and a half denarii of gold. They cannot yet part either the silver or the gold, but they hope to do it by a future better smelting process. The works in the mine are crippled. Wherever they found some ore they eagerly took it out, but left the work as soon as they broke off. Hence arose so many holes, that the work resembled rather those of rabbits than of miners. They have not yet any smelting furnaces of their own; accordingly they send their ores to the silver furnaces at *Csertes*.

In the afternoon I continued my way to *Nagyag*; and passed the other side of the *Maros* over high mountains, consisting of argillaceous rocks, mixed with mica and sherl, and covered with argillaceous slate. After three hours ride I reached the village *Nagyag*. It has given its name to the town, which is one hour and a half's journey higher up in the hills, since it was the nearest place when these gold-mines were discovered. I got here oxen to my carriage, because the little *Hungarian* horses, fit for plain countries, would not have answered these steep and high mountains. Towards night I arrived at *Sekeremb*, the proper name of the place commonly known under that of *Nagyag*. All around you see but forests, and in a valley some hundred houses, stamp-mills (*Pochwerke*) bing-places, some large washing houses

houses, the council-house, and a church. The situation of this place, and the coldness of the weather, are unfit for husbandry. The trade of the inhabitants is mining, and what belongs to it. The timbering of the mines, and the consumption of the inhabitants, have cleared the forests so much that the timber for the mines is to be fetched from abroad floated on the *Maros*, which passes at the foot of the mountains. The noblemen, as lords of the ground, have no objection against this clearing of their forests; nay, they keep great herds of goats, to prevent their growing up again. Every nobleman keeps on his ground an inn, to sell wine to the miners; and as the proprietors of the mines have engaged to discharge every month what their workmen may owe for wine, they have allowed for it the liberty to cut down in the forests whatever may be wanting for their mines and buildings.

The mountains are here entirely composed of our metallic rocks (*Saxum metalliferum*) which are covered with red argillaceous clay. The gold mine owes its discovery to accident.

A *Wallachian*, whose name was *Armenian John*, came to my father, then possessed of a rich silver mine at *Csertes*, telling him, that as he constantly observed a flame issuing from and playing upon a fissure in the *Nagyag* forest, he was of opinion,

that

that rich ores muſt be hid under ground. My father was fortunately adventurous enough to liſten to this poor man's tale; and accordingly he drove a gallery in the ground, which the *Wallachian* had pointed out. The work went on ſome years without any ſucceſs, and my father reſolved to give it up. However, he made a laſt drift towards the fiſſure, and there he hit the rich black and lamellated gold ores, which firſt were looked upon as iron glimmer, but appeared what really they are as ſoon as aſſayed by fire. This happy accident cauſed my father to purſue the work to the utmoſt of his power; accordingly he diſtributed ſome ſhares among his friends, and had the works carried on with regularity. Soon after they diſcovered, beſides the *Ergezebaw* and *the white fiſſure*, three other fiſſures in the hading ſide, and a ſoaring fiſſure, which, moſtly parallel among themſelves, run in the direction of the valley from ſouth to north, dipping from weſt to eaſt. Theſe veins break off as ſoon as they reach the red ſlate which covers the valleys. The cauſe is obvious to you, and gives me good hopes, that whenever we ſhall chace theſe veins under the plane, on which the ſlate is ſuperincumbent, their run will be uninterrupted, ſince then no cauſe remains to intercept it.

In the oppoſite mountain we have diſcovered another fiſſure, called *John Nepomuck*. It has a conſtant

constant run to the same point north, in which probably all these veins will cross and meet together.

The *John Nepomuck* vein has proved dead hitherto; however some small nests of ore have been hit already, and those permit us to hope that it will prove richer in the ascent of the mountain.

All these veins fall, and have been worked already to sixty fathoms. It has been observed that those, which towards the day or the turf were poor of silver and rich of gold, proved in the depth richer in silver and poorer in gold. The reverse happens with those that in the uppermost galleries yielded more silver and less gold.

Hitherto we have the advantage to run our ores by the galleries immediately to the bing places and stamp mills; and many years ago we begun (*erb stollen*) a draining gallery, which goes thirty fathoms depth under the present deepest sole; nay, the nature of the ground allows us to think of a still deeper gallery.

Our draining gallery is driven twelve fathoms in a bed of coarse blunt pebbles mixed with some clay. Had it been more indurated it would have proved a fine Breccia. Then followed the red shivery clay, through which we have

forced

forced our way 370 fathoms length. At prefent we work in fandftone, which grows harder as we go on, and gives good hopes that we fhall reach the micaceous clay rock, and in it after a twenty fathoms drift the veins which crofs it.

The mouldering and foul quality of thefe different ftrata has made our galleries expenfive and difficult.

To have at all events proper air conductors, and the requifite room for the water channel, it was to be twelve foot high; and the fides and roof were to be faftened by oak door ftocks, each a foot thick, and fet clofe together.

At the entrance of the gallery is a fan ventilator brifkly turning by a water-wheel in a clofe room, whence wooden air-pipes and conductors convey the air at the bottom of the drift, and caufe its requifite circulation. The conductors *(wetter-lotten)* confift of four boards nailed together, and made tight in the joints by a cement of clay, tile-duft, and tallow.

The workings are extremely regular. You would rejoice to fee in many places four, nay even five platforms one above another continually yielding the richeft ores; and it would certainly pleafe you to find in many long drifts a roof of three or four foot large veins, which enfures to the proprietors rich dividends in future. The miners

miners are not allowed to work in the richer veins, but near them in the hanging fide. Being nicely laid open that way, the fole is wiped clean, cloths are spread upon it, and then the vein is taken down in the prefence of an officer. This they do at the end of any working day, or at the end of the week; and prevents not only fpilling the ore among dead rocks but ftealing it too.

By the *Daniel-fhaft* the air is conducted into the deepeft drifts, and the conveying of the ores in the galleries promoted.

The vein rocks confift of red feld-fpath and white faponaceous quartz, *(fetter quartz.)*

The richer ores are lamellous, fplendent and black-grey; the lamellæ to be feparated from each other by a needle as thofe of mica. They may be cut and bent.

Another rich fpecies is finely woven into the fubftance of a bleak reddifh feld-fpath, refembling the arfenical white ore from *Saxony*; but the fire proves it to be native filver, of a yellowifh colour, on account of its mixture with gold.

Among the rich lamellous ores now and then occurs native filver mixed with gold.

Another rich fpecies is called by the miners *cotton ore*. It confifts of little native filvery gold grains, in a black gold mulm, fticking in an argillaceous matrix.

The other ores are likewise lamellated, but these lamellæ are but thinly sprinkled into their substance. Some are entirely similar to the scaly antimony and stain the fingers; others have a mixture of black lead lamellæ, which in closed vessels are unaffected by fire, but under the mufle yield a small corn of gold. In the midst of this ore is very often found a radiated crystallized, but commonly a scaly and plumose grey antimony. *Antimonium plumosum.* Red solid and crystallized arsenic, calx or orpiment, *calx arsenici, sulphure mixta, rubra. Cronstedt.* §. 241. n. 2. and fine grained cinnabar, are not uncommon. All these semi-metals brought to the couppelle leave signs, nay now and then small grains of gold.

The richer ores are in wooden troughs carried to the separating rooms, and there as nicely as possible separated by officers under oath.

The richest species contains from ninety to 340 ounces of silver in a hundred weight; and each mark silver yields 200, to 210 denari gold, that is to say, twelve or thirteen ounces gold, or two parts gold and one part silver. The rock separated from this ore, yields from fifteen to twenty ounces silver, and this from 160 to 170 denari gold per mark. The splittings and offal of this ore gives from twenty-five to thirty ounces silver, and this from 180 to 190 denarii gold.

The

The poorer ores are separated in the wash-works by iron sieves. The greater pieces, which do not pass the first sieve, and those that pass the first and second, are with hammers separated from the dead rocks. Those that pass through the third and fourth are taken care of by the sieve-masters, and the dust running through the remaining sieves is washed on the common hearths. These pyritous ores give two or three ounces silver, and this in the mark from seventy to 112 denarii gold. The separated dead rocks of this ore are stampt, and with the common ores pulverized and washed. They yield one or one $\frac{1}{2}$ ounce of silver, and this per mark from 100 to 130 denari gold.

Whatever care you may take with stamping and washing the richest *Nagyag* ores, the best microscope will not discover any flake of native gold. Counsellor *Scopoli* has chemically analysed them in his *Anno* iv. *Historico-Naturali*. Professor *Schreber* has given a translation of his treatise in his collection of Finance Tracts. You have read them, and you will have perhaps an opportunity to examine the constituent parts of this unique gold-ore and Mr. *Scopolis'* Essays.

The ores are assayed every month, and accordingly separated. The richer ones are pounded in iron mortars, sprinkled with water, put into sacks,

sacks, and together with the sprinkled ores and washings carried by horses over the mountains to the royal market at *Zalathna*, there to be assayed again by a royal assay-master, and to be paid accordingly to the proprietors.

The ores being watered they count at *Zalathna* three pound per hundred weight less. 2. They are charged smelting expences two florins a hundred weight. 3. And five per hundred fire loss for the gold as well as for the silver. After these previous deductions the mark gold is payed to the proprietors 300 florins; and the mark silver nineteen florins thirty cruizers.

Victuals being extremely dear at *Nagyag*, as being carried there by men or horses, the wages of the workmen are higher here than in many other places. These and the common mining expences amount from six to 10,000 florins a month. Nevertheless a dividend of eight, ten, nay of 20,000 florins is distributed every month among the proprietors; so that in twenty years time above four millions of florins gold and silver have been produced in this single place.

The proprietors have transferred upon her majesty the Empress-queen the principality or the right of regulating the mining works. Her majesty was possessed of sixteen shares or actions. However, the proprietors are in cases of importance

tance ſtill aſk'd their opinion. At preſent the works are under the direction of Mr. *Daniel Caſtellano*, her majeſties barmaſter. He is a miner of great experience, well acquainted with the nature of theſe mountains and ſucefsful in his undertakings. He is the firſt in *Tranſſylvania* who built regular ſtamp-mills, and demonſtrated the advantage of the *Hungarian* ſtamp and waſhing-mills to thoſe that objected, by clearing in a ſingle day 300 weight ſtampt ores with ſeven ſtamp and waſh-mills built at *Nagyag*. The ſcarcity of water has very often in dry weather put a ſtop to them; therefore the proprietors are building at preſent a great water reſervoir in a higher ground, in order to ſupply the mills during dry weather.

LETTER

## LETTER XII.

*Zalathna, July* 15, 1770.

*TRANSSYLVANIA* deserves to be examined by a naturalist endowed with a proper knowledge of minery. All the mountains of this beautiful country are full of signs of undiscovered metals. Had I not had a time prescribed, within which I was to return to *Shemniz*, I should not have left so soon a country which is so interesting to me, not only for being my native country but for being so rich in natural curiosities. It would certainly have made me rich amends for the pains of my enquiries.

The best and only reputable book on the Natural History of *Transsylvania* is, *Samuel Koleseri from Keres eer Auraria Romano Dacica.* Printed at *Hermanstadt* 1717. The author, a learned physician, and afterwards inspector-general of the *Transsylvanian* mines, has in this valuable book rather described the antiquity and florishing state of the *Dacian* mines during the reign of *Trajan*, than their productions and natural circumstances. With greater presumption but less learning a certain jesuit, P. *Fridwalzky*, attempted of late the

the Mineral History of *Transylvania*. His book has nothing to tempt you but the title, *Mineralogia Transylvaniæ*. The materials of this dull performance are a variety of good and false accounts compiled from able but generally ignorant miners, which the good-natured priest was unable to make use of; false denominations of minerals arising from want of knowledge; half a dozen pious tales, fit for the entertainment of old gossips; some authorities taken from *Kolefer*; absurd inductions, consequences and conjectures, arising from a thorough ignorance of chemical and mineralogical principles; and a good deal of civility and compliments to those gentlemen, who entertained *P. Fridwalzky* with hospitality in his excursions of mineralogical knight-errantry. Even the language, in which all these fine rareties are described, is such bad Latin, and so overdone with flourishes, that one is in want of the sense of *Œdipus* to guess that of *P. Fridwalzky*. I shall have perhaps an opportunity to converse with this highly celebrated *Transylvanian* mineralogist at *Clausenburg*; then I will tell you who he is, and whether perhaps in future times we may expect some amends for the singular raree show of his former performance. The mineral history of my noble country forces me often to a desire to make here a stay of some years, in order to satisfy my curiosity, and to hunt it over in the

remotest

remotest corners. At present I can only usefully employ the short stay which is allowed me; therefore I went the 13th from *Nagyag* to *Zalathna* here to examine the neighbouring mines. Our metallic rock *(Saxum metalliferum)* composes the mountains, over which the roughness of the roads allows no ride but on horseback. These argillaceous rocks are two hours journey beyond *Nagyag*, either entirely bare or covered with a reddish indurated shivery clay. Near *Barzcha*, a *Wallachian* village, rises a still higher calcareous mountain, superincumbent on the before-mentioned ground. It shews every where indications of copper ore, and some adventurers have worked for it without success. Near the village *Glut* the mountains slope into the plains and the limestone disappears, instead of which the red shivery clay is seen again. Being dissolved in its surface into a red mould, all the country about *Zalathna* seems to be red coloured. After a five hours ride I reached this place, which is at present, what it was in *Trajan*'s time, the seat of the upper court of mines. The many old inscriptions offering hereabout, and mentioning the *Procuratores Aurariarum Daciae* and the *Collegia Aurariorum*, established in these parts, make *Zalathna* extremely interesting for antiquarians. *Zamosci*, *Lazius*, *Kolefer* and *Fridwalzky* have compiled

compiled and published these inscriptions. This place is situated in a pleasant valley, intersected by the river *Ampoi*.

The *Wallachians* consider this town as the metropolis of their nation in *Transsylvania*, and repair to it in the market days. The greater part of the buildings are inhabited by mining officers.

The administration of the mines established here differs from that in the Bannat in this circumstance, that every society of actionists may as they please work their mines independant of the royal officers, under the condition however to deliver their gold and silver in the royal office at a fixed tarif of 300 florins a mark gold, and of 19 florins 30 cruziers the mark silver, with a deduction of five per 100 for fire loss. Societies, that have not got any dividends, have now and then allowed them a higher price.

The upper direction is subordinate to the *Transsylvanian* chamber of finances at *Hermanstadt*; but this in respect to the mines to the court chamber of the mines at *Vienna*.

The justice of the mines is independant of this direction, and decides the variances of the different societies and the miners. Besides there is a royal gold office, where at certain days the wash-gold of the *Wallachians* and gipsies is bought at a settled

tled price at two florins thirty cruizers per pifeth; three pifeth three denarii being equivalent to one ounce of gold. If the gold be already purified by mercury, it is bought fifteen cruizers dearer.

This office is a great advantage for the poor *Wallachians*, fince they are enabled by it to fell every week without impofition whatever be their fmall provifion. As commonly the gold-grains fold to this office are fcarce three or four denarii weight, it is impoffible to affay each feparately, and to pay them according to their proof and real value. Hence arifes the neceffity of an equal price, without any refpect to their different interiour value, and confequently an opportunity for the *Wallachians* to adulterate and heighten the weight of their gold by a mixture of filver filing, as foon as they are convinced of its fuperior purity. You fcarce would imagine that thefe fmall and trifling grains of the *Wallachians,* amount every year in the whole country from feven hundred to a thoufand weight of fine gold.

Not to lofe any time I haftened after dinner to the rich *Maria Loretto* gold-mine on the *Facebay* mountains, near *Zalathna* to the north. After half an hour's journey I reached the foot of the mountains, covered with argillaceous flate, fuperincumbent on grey horn-flate. The mountains rife gently; however, at firft fight of their
high

high elevation they seem to be steep and inaccessible. At an elevation of 150 fathoms the ancients have driven the *Sigismund* gallery. According to a traditional saying it was begun in the fifteenth century, under king *Sigismund*, and yielded then amazing treasures; but considering its having been cut through a solid hard hornstone, above 300 fathoms in length and six feet high, its sides being fair and even as stone-cutters work; and the length of time requisite to work out such a long way without blasting, one is rather inclined to look upon it as a work of *Roman* slaves condemned to the mines. Perhaps the works were only taken up again under king *Sigismund*. It is remarkable that this gallery is driven in a straight line and direction to the fissure; and hence it appears to demonstration,* that the *Romans* knew to apply geometry to subterraneous

---

* If it was proved to demonstration that the *Sigismund* gallery really is a *Roman* work: however, we need not be anxious about their subterraneous geometry. The *Grotta di Pausilippo*, the subterraneous aqueducts from *Lago Albano* near *Castel Gandolfo*, the stupendous *Cloaca maxima* at *Rome*, and many other of their buildings, are unquestionable evidences of their skill under-ground; and they might, as well as our common miners in *Derbyshire*, with a simple board, or rude *tabula prætoriana*, take up the subterraneous angles, and by that means, without any magnetical needle, hit under-ground whatever point they pleased. (Transl.)

works of this nature. In the roof of this gallery are some indications of former air-conductors, and these seem to have been considered in its height. The vein runs from south to north, and is crossed by the above gallery, which runs in west.

*Quartz* and *Hornstone* (*Petrosilex Cronstedt*, §. 62.) are the vein-rocks, in which offer the auriferous pyrites. These contain from two upwards to ten, twenty, forty, nay sometimes more ounces of gold. Besides an auriferous pyrites of four ounces value, they find a common grey clay, which yields some gold by washing.

This mine is not at present under the most fortunate circumstances. Superincumbent on it is grey clay and argillaceous slate, which probably seem to be the rocks of the *Maria* mine, situated somewhat higher to the west. I could not see it; since I was to examine the famous *Loretto* mine and then the return.

About fifty fathoms above *Sigismund* mine I met with the sandstone, which is the rock of *Loretto*, and accordingly is superincumbent on argillaceous slate. The gallery was driving in a finer yellowish sandstone. *Cos particulis distinctis.* Then followed a species of stone, which is called *backstone* and consists of blunted rocks ferruminated by common clay. *Breccia arenacea, Cronstt.* §. 275. Afterwards came grey hornstone,
*Petrosilex,*

*Petrofilex*, in which two veins, fourteen fathoms distant from and parallel to each other, are running. One is called the silver the other the gold vein. Both have smaller concomitant ramifications. The silver vein contains auriferous fallow ore, yielding up to eight ounces of silver per hundred, and its silver twenty and more denarii of gold in the mark; however, according to a common ill opinion against the silver mines in *Transsylvania*, it is not working. They are the more assiduous in the gold vein.

In the midst of the grey hornstone is a round cone or wedge of sandstone, as forced in from the turf. It is from two to three feet diameter. They have sunk in it a shaft eight fathoms deep. This sandstone consists of a variety of successive coarse and fine grained, grey and yellow horizontal compact sandstone beds, of a different thickness; some being not above one inch, some above a foot in thickness. Every bed has a particular gold mixture, which arises from the auriferous pyrites sprinkled more or less in these sandstone strata. One of the inferior beds for example yields four ounces, the superincumbent one 100 ounces, another two ounces, another fifty, and another still superior to those 200 ounces gold per hundred weight. As far as I may judge by the samples which I have taken with me, it seemed to me that the finer grained fandftones are the richeft. However, the infpector and chief proprietor, Mr. *Weiſſe*, aſſured me, that this obſervation is not general.

Another circumſtance in this problematic mountain puzzled me ſtill more. I obſerved in the grey hornſtone a great number of round holes, three or four inches deep. At firſt I conſidered them as the remains of blaſting-holes; but enquiring after the reaſon of theſe ſo numerous miſcarried blaſtings, I was aſſured that theſe holes are natural to the rock, and that each of them naturally contains a blunted pebble. Really I found in the greater part of them blunted pieces of ſilex, or of indurated clay, which ſeem to have been hardened and blunted by rolling before they were incloſed in this hornſtone. I here frankly confeſs my ignorance. It is impoſſible for me to explain the origin of this paradoxical mountain, or with the utmoſt ſtretchings of my fancy to create a tolerable hypotheſis. See whether you can aſſiſt me, or whether your obſervations in other parts of *Europe* enable you to explain this phænomenon.

This mine I fear will not be long working; becauſe, whatever be its riches, it ſeems to be in a diminiſhing ſtate, as the veins in the hornſtone begin to break off; and after the draining of the
little

little stockwork of auriferous sandstone no great hopes seem to be left, nor any probable chance to make new valuable discoveries within the small compass of a ground which is at the top of the mountain. This is the more to be apprehended, as, by an unaccountable neglect, they have not thought in so rich a mine of a deep gallery or drift, which being driven a-cross the whole mountain, would most certainly have laid open, whether or no there are other veins and riches under-ground. They have but of late begun to drive a gallery under the pit, which will enable them to sink it ten fathoms deeper.

The ores of the *Face bay* mountain are remarkable phænomena for mineralogists. Common pyrites of no promising appearance contain from two to 600, nay to 900 ounces gold. On some pyrites the gold appears in a metallic form; on others it is sprinkled as *Spanish* snuff (*Brunnich's new edition of Cronstedt's Mineralogy*) on many others no gold is to be discovered by the strongest microscopes. This species may be stampt to the finest powder, and no washing will produce the least suspicion of any gold-dust.

The workmen know their inner value at first sight, and they are so very skilful in separating the ores according to their value, that the assay-master of the proprietors makes no other sorting,

but

but takes of each fort a general proof; whereupon they are delivered to the royal furnaces to be affayed again by a royal officer, and to be paid their thus ftated value.

Any bit of thefe pyrites brought under the mufle, or to any other fire, the gold appears prefently fweating out on the furface in little globular grains. The fame happens with the gold-ores from *Nagyag*. This circumftance has caufed fome mineralogifts to fuppofe, that the conftituent parts of the gold, hid in the fubftance of the ore, unite by the fire, and that the wanting parts, probably the requifite phlogifton, being added by the heat, facilitate the operation.

The ftamp-ores, confifting commonly of fine pyrites, fprinkled in the hornftone, and digging next to the veins, are roafted to make them brittle. But that caufes a fenfible lofs to the proprietors, fince the fire expelling the gold from the fweating hornftone in the form of a fine duft, this precious powder is too eafily carried away by the waters of the ftamp trough. This bad practice is entirely owing to the nature of their ftamp-mills. Their pound-rammers or grind-peftels *(pochftempel)* being armed inftead of iron *(poch-eifen)* with black or grey hornftones, cut into the requifite form, the roafting of their hornftones is neceffary in order to make them

them brittle. In general, these stamp-mills are extremely imperfect. They are uncovered; accordingly every rain and flood carries away a good deal of their pulverized ores. Their water wheels are too small; hence arises a great loss of water. The dust channels are of the same width, and have no inclination at all; hence no such thing as a separating the richer from the poorer dust. I have shown all these deficiencies to the mill-master, and to convince him of the bad consequences of his stamping, I caused some sands from the first and last channel, and even from the flood or the brook that runs by, to be washed in his presence, which clearly shewed that all these sands are equally rich, and that the gold in the brook-sand is downright loss to the proprietors. But for such people old customary practices are above reason and conviction.

Four hours journey from *Zalathna* to the east is *Abrud-banya*, the former seat of the upper mine-council. Red, and now and then grey schistous clay, covering our metallic rocks, appeared in all the mountains about *Abrud-banya*. The most remarkable ones are *Igric, Csetate, Boylor, Korna, Orla, Kirnizel, and Kirnik*. It is impossible for you to form an idea of their workings. The whole *Kirnik* mountain is from every part and side perforated with many hundred galleries, which do not penetrate above some fathoms into its inner parts. Though this seems to be extremely bungling and aukward, it is less so in respect to the nature of its fissures and veins. Its numerous gold-veins are thin and short. The adventurers have allowed a field of three fathoms in the hanging and as much in the hading side. They commonly begin working on a vertical fissure, *(seiger kluft.)* which they work for six or seven fathoms; then it begins to dip and to be inclined, and insensibly it turns flat; that is, according to the nature of this mountain, it turns rich and yields native gold. But it continues scarce above two fathoms flat and noble; since suddenly it turns again and breaks off. The miner knows by experience, that now he is at the end of his hopes on this fissure; therefore he drops it for another, or searches in the old drifts, till he meets with some worthy remains of the old man. Hence that innumerable quantity of galleries or holes.

A *Roman* inscription, decorated with miner-instruments like ours, and found near this place, is an evidence that the *Romans* worked here.

Now and then extremely rich and showy ores are found; however, the proprietors, being often at work during several weeks without any success, are generally poor, and think themselves happy by
getting

getting a week the scanty revenue of three or one ½ florin. The greater part of the inhabitans have no trade but this mining. The father is commonly buried in his mine; the son carries the ores to the mill; and the women take care of their stamping. The children gather the sands and the mud, which by rains is carried in the valley, bring it to the mills, and it generally yields some gold.

To have this singular *Kirnik* mountain the better examined, a gallery 300 fathoms in length is driven under it at her majesty's expence; but it has crossed only some deaf veins, has freed the workers from the day waters, and is left at present to their further disposition. The valley in which the stamp-ores are preparing is called *Voros Patak*. I do not exagerate in telling you that there are in it above 300 stamp-mills, which set at work make a noise so as to be heard at an hour's distance. But they are as the common stamp and wash-works of the gipsies, without any covering, and with a single sand-channel. Instead of the stamp-iron they use here a grey hornstone from *Korosbanya*, which *P. Fridwalzky* is pleased to call calcedony. I cannot be convinced of the assertion of the mine officers, that they stamp and wash here without any loss; and you will rather agree with me, when I tell you, that the inhabitants of a neighbouring village, called *Kerpenes*, live entirely upon the neglect of those at *Abrud-banya*. They dig holes, in which they convey the same brook which runs through *Voros-Patak* valley, and another which drives the stamp and wash-mills at *Bucsum*. The sands carried in these cisterns are auriferous, and pay richly the pains of a new washing at *Kerpenes*.

There are several other gold-mines in this neighbourhood as near *Bucsum* near *Abrud-Zeller* and in mount *Volkci*. They consist generally of auriferous quartz-veins, left by the ancients, and now worked over again by the *Wallachians*.

Near *Zalathna*, in the *Barsa* and *Rusina* mountains, are *Peter Paul*, the *Three Kings*, the *Saints*, and some others. A great many more have been abandoned; and several of them yield auriferous lead-ore, nay, some native gold, but to no great advantage.

I cannot pass silently over the two mercurial mines at *Zalathna*. The first is to the north of this place, at an hour's journey from *Dumbrava*. The cinnabar ore (*cinabaris solida, textura squammosa, Sqammis minimis*) breaks here in a vein, and in a matrix of quartz and sparr, inclosed in argillaceous black slate and sandstone. The vein runs from north to south, but is leaping, often a fathom thick, often compressed and deaf. The second is to the south in the *Baboja* mountain. Its ore is granulated

granulated cinnabar, digging in a vein, which runs in limeſtone. Probably the ancients dug here great quantities of cinnabar. At preſent the *Wallachians* ſeek here after the remains of the old man. Some ſocieties however have united of late to undertake regular workings at *Baboja*. The mercurial ores are delivered to the furnaces at *Zalathna*, but the annual amount of clear quickſilver is not above three tons. They clear it by common diſtillation in retorts and alembics filled with water. If they told me truth, there is of late built at *Kisfalu*, near *Clauſenburg*, a ſublimating furnace for preparing ſublimat-mercury

## LETTER XIII.

*Nagyag, July,* 1770.

THE first mining place, which I met with the 16th on my excursion to the mines west from *Nagyag*, was *Csertes*. My father had here thirty years ago a rich silver-mine, called *Trinity*, which yielded a good revenue. But the situation of the ground making a deeper draining gallery impossible, and the want of sufficient water for the pumping-engines forced us to give it up. The mountains to *Csertes* consist of metallic-rock, covered with common argillaceous slate. But the *Bogaja* mountain, in which the above mine was working, consists of sound compact hornstone; the veins, which cross it, are very rich, and yield auriferous glass-ore, closely woven in the substance of the hornstone. The rocks are so remarkably hard, that even with blasting the works advance but little. P. *Fridwalzky* advised therefore in his Mineralogy to soften these refractory rocks by bacon suspended and set on fire before the drifts. O! *Sancta simplicitas!*

A new

A new society of adventurers has at present undertaken to drain this mine by a new gallery. But as it is to be driven a long way under ground, and neverthelefs cannot go any confiderable depth, I am apprehenfive that, though the drained part of the vein fhould confift of the richeft ores, the expences will hardly be cleared.

In the adjacent hills are fome gold-works, which in former times have given confiderable dividends. Thefe are not in hornftone (*petrofilex*) but in metallic rock.

The furnaces near *Cfertes* are employed with the fmeltings of the neighbouring focieties. Want of water makes them often inactive.

The *Fourage* mountains near *Cfertes* have had in former times many mines. For the greater part they are given up. The inhabitants of thofe parts aflured me, that pieces of native gold, not lamellous but found as glafs-ore, had been found there.

I went from thence to *Topliza*. The mountains confift there likewife of that grey argillaceous rock, mixed with mica, fherl or quarz grains, which I have prefumed to call metallic rock (*Saxum metalliferum.*) They are covered with argillaceous flate. The veins are commonly a quartzous auriferous ftone, and conftantly running from fouth to north. Such is the vein of the *Nepomuck*,

*muck*, *Martinus*, *Rochus*, *Archduke Peter*, *S. Joseph*, *Mary's Annunciation*, *Florian*, *Francis de Paula*, and the *Holy Crofs* in the *Magura* mountains. I was told that in *Nepomuck* native gold had been found immediately under the turf. Some of thefe fiffures yield, befides their native gold, fine auriferous red filver ore.

In the *Matfhire* mountains they drive at prefent a gallery, in order to drain fome old works; and in the *Fifher* and other adjacent mountains are fome gold-works, which now and then yield confiderable pieces of native gold, but very often are dead and deaf.

In thefe *Toplitza* mountains the gold is often found in lead-veins. The fame happens in the more wefterly mines at *Fuezes*, whofe rocks in the *Malula* hills are entirely correfpondent with thofe at *Topliza*. Near *Fuezes* I found a grey loofe marle flate, inftead of the common argillaceous fhiftus, fuperincumbent on the metallic rock. Being diffolved by the air, and confidered as common clay, it had been employed to line the inclofure of a water refervoir. The water diffolved the lining, and the refervoir wafhed its enclofure away. Without enquiring the reafon, the fame marle was made ufe of again of late, and laft fpring the fame accident happened. At laft they were fenfible of the blunder.

The

The confequence of this double unhappy expenfive accident would prove very happy, if people would learn thence to conclude, that a mining officer fhould at leaft be acquainted with the common foffils and their qualities.

In *Clemens* near *Fuezes* there is native gold in felenite, or *gypfo fpatofo albo pellucido*.

On the oppofite fide of the *Malula* hills is *Trfztyan*, a place greatly renowned for its rich gold veins, and the magnificent fhowy pieces of native gold which are found here every day.

As they have received in *Tranffylvania* a principle, that native gold is to be found only immediately under the furface of the horizon, I was highly defirous to examine this mine, hoping to find perhaps fome arguments againft that opinion, fince for a very long time it has been worked to great advantage, and has produced an uncommon quantity of gold; which feems to me unaccountable, if the vein did not dip under the horizon. But the proprietor, Count *Stephan Gyulai*, fcarce allows any imperial mine-officer to vifit his mine; and all the works, being fuperintended by a *Wallachian*, are fo barbaroufly bungling, that a man muft be a *Wallachian* to hazard his life for his curiofity, and to flip down in the fhafts wherein no fuch thing as a ladder or other proper affiftance is to be met with.

Therefore

Therefore I was confined to examine the nature of the mountain, which confifts of the fame rocks as the *Fuezes* hills.

Better management would improve this noble gold-mine to a greater benefit for the proprietor. Being worked only by *Wallachians*, who never neglect any opportunity to pilfer, a good deal of the finer gold-ores may poffibly be concealed by the workmen. Some years ago I faw myfelf in the market place at *Deva* a miner from *Trfztyan* publicly felling fine famples of gold-ore. Though of late fome fevere ordonnances are publifhed againft this illicit thievery and fample felling, fince it lowers the benefits of the royal gold office, it will however be very difficult to prevent the lofs of the proprietors, becaufe any miner may eafily find an opportunity to fell his ftolen goods to the *Corfars*. This fort of people ramble about in the remoter mines, buying from the proprietors their little provifions of ftampt gold-ores, which by themfelves would not bear the carriage to *Zalathna*. Having gathered their full load, they deliver and fell them to the royal office. This trade feems to be an advantage to the royal office, as well as to the poorer adventurers and proprietors, but it degenerates too eafily into a commerce which proves pernicious to the richer works, as the *Corfars* have certainly no objection to purchafe the ftolen ores at a cheap price, and fufficient
skill

skill to pound and to mix them with the stampt ores, which they are allowed to buy and to sell.

At night I reached *Boicza*. The mountains hereabout connect with those, which at my arrival in *Transylvania* I found stretching on both sides of the *Maros* river. In general they consist from this place to *Deva* of metallic rock, covered with limestone, slate or sand. Some hills near *Boicza* are destitute of veins, consisting of blunted rocks, ferruminated by an argillaceous cement and resembling *Breccia*. The royal mine is working in a variety of metallic rock, differing from the common species by large feldspath pieces sprinked in its substance. The uppermost or *Anna* gallery was driven in limestone, which is superincumbent on the metallic rock, and covers large valleys; but the deeper gallery runs in sandstone till it reaches the argillaceous rocks. The veins and fissures are blendish-leadglance, containing some gold and silver. I have some samples with gold immediately sticking on the blende and the lead glance. At the tenth fathom of the deeper gallery I found an argillaceous fissure nearly vertical, and in it a great number of blunted oval transparent calcareous sparr-bullets with opaque milkwhite stripes, resembling those of onyx.

Limestone (*calcarius albus particulis granulatis minimis*) is hereabout a detested stone among the miners,

miners, since it cuts their veins or fissures. This however should not disrecommend the calcareous stones; since, according to the theory of superincumbent mountains, it is of a more modern origin, and deposited in the valleys in which in former times the veins and fissures baffetted out. The blendish lead ores contain here commonly three ounces of silver, and the silver sixteen denarii of gold per mark. The stamp-ores are henceforth to be prepared in three stamp mills, which are at present building after the model of those at *Shemniz*. They stand on a sloping ground, one above the other; pulverized ores to be washed in a great washing-house at the foot of the mountain, in which ten plane hearths will be set at work. A hundred weight of stamp ore gives eight pounds of metallic powder, the upper sort yielding six the lower two ounces of silver. The mark of silver contains sixteen denarii or one ounce of gold.

It was impossible for me to visit the many other mines at *Boicza*; but I have got some of their ores as well as those of several still working mines in *Transsylvania*, and these you will please to take notice of by this following catalogue:

Auriferous pyrites in bluish clay; *argilla communi plastica cærulescente*, from *Herzigen* near *Boicza*. They do not work here but on the old man.

Auriferous

Auriferous pyrites in black hornſtone from *Ginel* near *Boicza*. Native gold to be found here in the ſame matrix.

Native hair-ſilver on lead-glance from the ſame place. The rock ſticking to this ſample proves to me that the *Ginel* mountains are metallic rock.

Auriferous quartz from the old works near *Ruda* and *Kriezur*.

Native gold in calcareous ſpar from *Staniza*.

Native gold in ſtarry radiate antimony. From the ſame place.

Native gold in grey ſcaly cobalt (*ſcherben kobolt*). I aſſayed this cobalt, and had a gold grain left on the capell. The works at *Staniza* are in the *Jeſuina* and *Dimbul* mountains.

Auriferous pyrites in indurated ſhivery black clay from the croſs gallery at *Cajonel*.

Auriferous pyrites on quartz from *Gothelf* gallery at *Cojonel*.

Auriferous blende from the ſame place.

Auriferous lead-glance in hardened white clay from *Kiſbanya*, where at preſent a company has united for working the lead-veins of that place.

Auriferous red ſilver ore on quartz from *Trajka* near *Trſztyan*. The vein runs in metallic rock.

Lead-glance in quartz from *Offen banya* near *Zalathna*. Of late a company has united to work the large lead-veins which have been diſcovered

K there

there. The old works are of the same kind, it is not known for what reason they were given up many years ago. *Fridwalzky's* pretended reason is fabulous. The nature of the rocks, that is to say, argillaceous clay, superincumbent on our metallic rock; the width of the veins; the great number of old bing-places; and the marks of thirty furnaces hereabout, have co-operated to raise the company into high spirits, and to undertake the works with activity.

I got neither samples nor creditable accounts from the other *Transsylvanian* mines. However, to make my accounts of the known gold-works as compleat as possible, I will give you their names, taken from P. *Fridwalzky's Mineralogy.*

At *Nagy-Almas*, to the west from *Zalathna*, in the *Rudile Baba* and *Petrasack* mountains, gold is found in antimony, and in a species of stone which is unknown to me, and which P. *Fridwalzky* calls *spathor pyrites*.

At *Pojana* in the *Vertes* mountains, in the same tract of land, the *Wallachians* hunt after stampores left in the old works. In the year 1742 a blueish grey quartz vein with native gold was discovered by rain in the *Dimbul Kupiatra* mountain. It had its direction from east to west; but as in general the *Transsylvanian* gold veins are very inconstant and short, so proved this likewise.

likewise. After a short run it shifted its direction, turned and broke off. I guess from *P. Fridwalzky's* obscure description that it was in hornstone.

At *Porkure* in the *Csetras* mountains the *Wallachians* hunt after stamp-ores; which they afterwards stamp and wash at their account. In the *Vallkurethe* mountains marks of gold-mines have appeared likewise.

At *Kirosbanya*, a mining place in the bailliwick of *Wissenburg* and the *Mayura* mountains, a vein two fathoms thick is said to be interrupted by small auriferous fissures.

I reserve the *Transsylvanian* gold-washings for the description of my journey to *Upper Hungary*; and desire you to return with me to *Nagyag* for the sake of some iron and lead-mines on the other side of the *Maros*. The most remarkable mining place in the *Hunyad Comitat* is three hours journey from *Vaida Hunyad* to the east, near a village called *Gyalter*. It produces a good deal of iron. The iron ores are found here, as in many other *Hungarian* iron mines in nests, or stocks six or eight fathoms large, which have but an irregular and uncertain direction to the south, and do not sink into any considerable depth. The *Kropilela* mountains, which contain these ores, consist of grey and brown argillaceous slate. The ores consist of red and brown iron ocher, in which sometimes button-ore is wrapt up, covered with feather-like iron cryftals, as the button-ore in the *hulf-gottes* mine at *Platten* in *Bohemia*. The workmen call this button-ore *iron-flowers*. P. *Fridwalzky* is a good deal wifer; he calls it antimony.

The fmeltings have nothing particular, being done in a fort of fmall high furnaces; and the iron is beaten into bars in feveral hammer-works along the *Cferna*. The *Wallachians* and gipfies are in general blackfmiths and iron manufacturers. They ufe fmall and low furnaces, and blow the fire by portable bellows made of bucks fkin. Their conftruction is very fimple; confifting of a fimple fewing of the fkin of an iron air pipe fixed in the neck, and of two wooden handles, fixed to the fkin that covered the feet. The antiquity of thefe iron-works appears by an infcription found near *Oftrow* and fpeaking of a *Collegium fabrorum*. Perhaps even the denomination of the *Porta ferrea*, or the pafs on the limits of *Turkey* is hence derived. This remark and conjecture is entirely P. *Fridwalzky's*.

Nearer towards the *Maros*, and a village called *Kifmunes*, have been found fome lead-veins in argillaceous flate, which of late have been undertaken by a private company. On the road I found calcareous hills, filled with a great variety of turbinites

turbinites and other marine shells. These hills on the *Maros* are to be considered as the foot of the high mountains, which run by the *Haczeka* valley, uniting afterwards with the high *Granite*-mountains, between *Transylvania* and the *Moldaw*.

There is hereabout but a single natural curiosity remaining which I cannot leave unnoticed. Near the door of the deep *Josephi* gallery at *Nagyag* I have found an hill about thirteen fathoms high, consisting of an innumerable quantity of regular pieces of metallic rock. They are flat, and about a foot in thickness. This hill cannot possibly have been heaped together by human hands, nor is there any old mine to countenance this opinion. Besides, these shivery stones are exactly fitted and joined one to another; are not mixed at all with any other species of stone, and speak at first sight that by some accidental cause they are split and cracked into so many regular fissures and fragments. But what accidental cause? A concussion subsequent to the exsiccation of the rock seems to be a very probable one; but I have better reasons to consider these stones as volcanic productions. They are of a coarser texture than the common metallic rock at *Nagyag*, and are sonorous. I know that mineralogists will start many objections against this opinion, as in

a great

a great diftance no exftinct volcanos are to be met with. Therefore I defire you would examine this fhivery ftone, and a fimilar fpecies from the volcanic *Euganean* hills near *Padua*. Samples of them you will find among the *Tranffylvanian* foffils, which I have collected for you.

LETTER

## LETTER. XIV.

Near *Foldwinz, June* 24, 1770.

FATIGUED by the heat of the day I arrived here, where, except some grass for the horses, no accomodation is to be had for the passengers. Unable to swallow the sour wine of my poor landlord, I drank a glass of water. I think of you, and write to you an abstract of my journey from *Nagyag*. I left that place yesterday. When I had passed the metallic rock-mountains, which are covered with slate, I got into a plain, now and then interrupted by argillaceous slate-hills; then I reached the *Maros* to the right. To the left near *Bobolna* we had mountains of ferruminated clay and pebbles, as those near *Boicza*. They seem to have been heaped and washed together by the *Maros*, since their sloping to the river is less indurated, and a piece of tile ferruminated with other pebbles, which I found near the road, is a good evidence of their successive and modern origin. The hills of the same nature, which are more to the north, and those at *Boicza*, which are likewise on this side of the *Maros*, had probably

bably the same origin, and accordingly this river seems to have shifted its bed from south to north and to shift it ever more since the plains on the other side do not confine it within constant bounds. Hereabout, but on the other side of the river, are the *Olapian* plains, times immemorial famous for their gold-washings.

The surface of these plains consists of sands and pebbles. After the removal of the turf and of the vegetable mould, this sandy stratum, two fathoms thick, appears. Times out of mind it has been dug out, and yielded gold by washing. It is superincumbent on argillaceous slate, which is destitute of gold. To the east and west it is surrounded by hills, which in the south connect with those that divide *Wallachia* from *Transsylvania*. To the north it is adjacent to the *Maros*. It seems to owe its origin to some inundation, perhaps even to the *Maros*. The gold cannot possibly be considered as produced in this plain; it is probably washed by the rains from the adjacent gold-fissures and deposited with the sands. This conjecture gains some credit, because gold has been of late washed with success in such places, which many years ago had been worked and washed out already.

In the night I reached *Carlsburg*, a fine fortress. It is my birth place, and I had here my education

till

till the eighth year of my age. It is pleasantly
situated in a plain, surrounded by argillaceous
slate and limestone hills. I met here with an
*Hungarian* nobleman, who was very well acquainted with the gold-washings, especially those
in *Transsylvania*. He gave me the following accounts, which I communicate to you, since I
shall have no opportunity to examine these washings myself. All the *Transsylvanian* rivers and
brooks, nay even the sudden and momenta-
ry rain and mountain brooks are auriferous.
But the *Aranyos* river is by far the noblest of all
in that respect, and is compared therefore by the
*Transsylvanian* Historians to the *Tagus* and *Pactolus*.
The gold-washers are either *Wallachians* or gipsies.
These gipsies are not in the least resembling those
idle and lazy ones in *Hungary*. They are a laborious people, and honestly active for their liveli-
hood. Some are strolling fiddlers and musicians;
some blacksmiths; others deal in cattle and
horses; and the greater part has the gold-washing
business. They pay their poll-tax every year with
some hundred piseths of gold; and sell a good deal
to the royal offices. They have great skill in
finding and tracing out those places where gold-
washing is attended with success. Their tools
consist of a board two or three feet wide, and four
or five feet long, commonly edged on both sides

with

with a wooden brim. Woollen cloths are spread on it, and the sands, poured with water upon it, leave the finer and heavier sediments in these cloths, which afterwards are washed in a great water cask, and then by the common severing trough separated from the gold. If the sands be mixed with coarser gravel, the board has deeper cross-furrows, in order to stop these coarser stones, and to examine them for gold, which often is found visibly sprinkled in their substance. Such is their manipulation at *Topansalva*, near *Abrubanya*, and all along the *Aranyos*.

Another practice common in *Transsylvania* is to dig pits, and to catch and stop in them the sand and gravel carriage of the brooks, in order to sever them from the gold-dust or ores. I have observed the same at *Kerperes*; and it is practised at *Zalathna* on the *Ampoi* river, near the old mercurial mines, whose crrriages contain a good deal of mercurial ore.

The third method is to fetch the auriferous vein rocks from the old mines, and to clear them from the gold both by pounding and washing. This method is generally practised where plenty of water allows it.

This morning I travelled over a fine cultivated plain to *Enged*. Here is a *Calvinist* academy, and some schools of that religion. The adjacent

jacent hills calcareous; the whole place built of pale yellow sandstone ferruminated with lime. The sandstone, filled with plenty of petrified shells, and dug in the hills, which continued behind *Enged* to *Foldwinz*, and to the very place whence I write you these lines.

LETTER

## LETTER. XV.

*Claufenburg, July* 28, 1770

UNDER the moſt tremendous thunder and rain-ſtorm, which I ever beheld in my life, I arrived the 24th at midnight at *Torda.* Behind the place whence I wrote you my laſt the mountains are ever aſcending. From the top, whence I had a ſight of the baſon of *Torda,* I diſcovered a great many hillocks, ſuperincumbent on this high elevated ground. Though I could not examine their nature, they conſiſt probably of the ſame grey limeſtone which covers the valley, and ſtretches to the *Aranyos.* *Torda* is on the other ſide of this auriferous river; the ſalt works half an hour's diſtance from the town on an argillaceous ſlate hill, which is ſurrounded by a great many little hillocks, ſaid to be calcareous, and proving at firſt ſight that they owe their origin, together with the ſalt rock-mines, to former ſeas. The whole plain on this high ground contains ſolid tranſparent *ſal-gemmæ,* probably ſuperincumbent on ſchiſtus. I could not examine it myſelf, and the miners were too unconcerned to know

whether

whether their salt-rocks have a basis of clay or of limestone, or of any thing at all. The turf and vegetable mould which covers them shews commonly white efflorescences of kitchen salt, which exposed to the sun tinge the whole surface with white. They are to be ascribed either to the evaporations of the ground, or to the rain water running over the salt-bing places.

There are several mines or shafts sunk in the same salt-rock stratum. Their construction is particular. As soon as a shaft is sunk and timbered in the upper earth bed, which is commonly from three to six fathoms thickness, they reach the salt-rock, and work down in it a conical pit, so that all the miners are employed on the same sole. The number of the workmen is increased in the same proportion as the cone widens. At first sight one might believe the whole salt stock through its thickness of thirty or forty fathoms consisting but of a single stratum; but on nearer examination it appears to consist of many accumulated parallel beds, of one or two foot thickness, which are either horizontal or undulating and separated from each other by a thin layer of clay scarce the thickness of half a line. This natural separation facilitates the breaking of the salt-rock, and is the more an advantage to the workmen, as they are paid only for those pieces which are eighty pounds, the

smaller

smaller ones being thrown on the bing places among the rubbish. For a piece of the requisite weight they are paid half a groat; and whatever has the requisite bigness is carried to *Carlsburg*, whence it goes by the *Maros Hungary*. I descended with five gentlemen into the *Theresia* mine. Put all together in a wide sack of rope net-work, and let down in the drawing-shaft; the sack contracted by our weight so much that only our heads peep'd out. A miner was above our heads clinging to the rope which let us down, to avoid our clashing against the sides of the narrow shaft. This shaft is sunk ten fathoms through the hardened clay, superincumbent on the salt-stock; and a small gallery is driven on the surface of the salt beds to bring away the waters soaking through the earth and clay roof, in order to prevent their falling down into the works. Besides, there is a smaller shaft for descending and ascending of the workmen; but the works below being of a conical form, the ladders cannot possibly be fixed to the sides; accordingly they are fastened by iron cramps or ropes one to another, and hang free and swinging in the midst of the deep and wide cavern below. However, the workmen do not care, and being used to it, ascend and descend these thirty or forty fathoms on swinging ladders as unconcernedly and

and nimbly as other miners in the moſt regular ſhafts. When we reached the opening of the ſalt-ſtock we hung free in our ſack, and I ſaw with pleaſure below me, in the depth of thirty-three fathoms, the many lamps of the workmen. I found in the mine the director of theſe ſalt-works, who had the politeneſs to ſhew and to explain whatever was worth ſeeing. I was agreeably ſurprized by a burning bundle of ſtraw, dropt down the ſhaft. It illumined the whole cavern, ſhewed me its conical form, unſupported by any timber, and made me diſtinguiſh the undulating form of the ſalt-beds. The light was from every ſide reflected by the whiteneſs and brightneſs of the ſalt-rocks. I have examined the clayiſh earth, which ſeparates their ſtrata. It has a ſouriſh taſte, but an offenſive diſagreeable ſmell, like that of rotten cheeſe. It is tough like clay. The preſent ſole of this work is fourteen fathoms diameter. The *Coloſer* mine, of the ſame nature and form, is ſixty fathoms depth, and fifty fathoms diameter.

Many old ſalt-pits are given up long ago. Theſe are entirely filled with water, which they uſe hereabout for bathing. I was preſented here with ſome tranſparent pieces of ſalt-rock with incloſed water drops; another piece contains moſs. I got here likewiſe a great quantity of the *lapis numiſmalis*

*malis Transsylvaniæ,* described if I am not wrong by *Bruckmann.* Gyps and alabaster is very common hereabout. Pray tell me what is the reason that these stones are for ever to be found in the salt-works? Might not the saline acid perhaps be changed into vitriolic acid, and thus gyps be produced? I have gyps from the salt-mines in *Upper-Austria,* and from the *Marmaros,* where it is found between the salt-beds.

After having satisfied my curiosity under ground, I saw the immense piles of rejected salt-rocks which are under the prescribed size. They are kept for no use; and severe penal laws forbid even the poor to make use of them. The reasons which they gave me for this unaccountable squandering away of so useful a substance, were as follows: That to prevent smuggling pieces of the same weight were given to the carriers, and that the abundance of the *Transsylvania* salt-mines did not seem to require any grudging, or sending the smaller salt-pieces in sacks or casks. This might do, if the world was only to last but a thousand years more; but good public œconomy takes care of the latest posterity, and disapproves any arbitrary and unnecessary destruction of an useful mineral as inhuman extravagance. Many hundred millions weight of rejected salt are thus exposed to dissolving rain and snows; and what vast quantities are spent in in a similar way near and in the *Vizakna, Kolofer, Szekes, Deefe* and *Paraite* salt-mines? Though P. Fridwalzky considers the *Transylvanian* salt-mines as inexhaustible stocks, he could not digest this cruel wasting of useful materials, and he had in this *Mineralogia Daciæ*, p. 171, the fantastical idea, to mix these immense salt-masses with tartarus, and thus to change them into nitre, in order to make sublimate. However, there might be some chemical use of this wasting-salt. Might not saline acid be properly joined with urinous substances, every where to be had, and thus sal-ammoniac be produced? You know of the sal-ammoniac-manufactory of Mrs. *Gravenhorst* at *Bronfwic*. Might we not possibly expect a greater advantage from our superfluous salt-rock pieces, than these gentlemen can get by their scanty and aqueous brine?

*Claufenburg*, two hours ride distant from *Torda*, is divided from this last place by a high mountain, consisting of argillaceous slate, and offering from the middle to the top a great many globular stones, three, four, and five foot thick. They consist of yellow sandstone, cemented with lime and filled with petrifactions, such as I found stratified near *Foldwinz*. Some of these stone-bullets are closely sticking together; which supposes their having been connected when in a soft state.

This mountain ftands completely infulated, and there is no higher mountain round about, from whence thefe bullets might have been rolled to this. Accordingly they feem to have been carried thither by the floods of the fea, whilft it covered this country. At the foot of this mountain lies *Claufenburg*, one of the fineft and moft populous cities of *Tranffylvania*. The *Roman* monuments, quoted by *P. Fridwalzky*, and found hereabout, prove that in former times it was a *Roman* colony. The houfes and even the walls, which inclofe this place, are built of grey or yellow limeftone, filled with fand and petrifactions. In general the country abounds in them; however, I have not found any fcarce fpecies. The defire to converfe with a mineralogift, or rather my curiofity, prompted me to vifit *P. Fridwalzky*, living here in the college of the jefuits. His rooms are filled with ill chofen ftones, minerals and petrifactions confufedly piled up. They bear the mark of their poffeffor's confined and unconnected fcience. He is really a very induftrious laborious man, but he has got together fuch confufed ideas of natural-hiftory, that I am apprehenfive he never will be able to bring them into any tolerable order, or to diftinguifh the true from the falfe ones. This is rather owing to want of proper inftruction and ufeful books, than to a deficiency

deficiency of his application or capacity. The desire to be useful led him to compile a *Mineral History of Transsylvania*; ignorant what science and experience is requisite to such a task. He supposed to have discharged his duty by giving the names of the mines, the length of the drifts, the depth of the pits, by compiling some accounts of mine-officers, a couple of charters and other monuments; and then, after all, telling his countrymen that this is a *Mineralogia Transsylvaniæ*. No such thing as proper descriptions or denominations of the rocks, and their local connexion and variety, which would have been material. At present there is scarce any hope of his improvement. The good-natured *Transsylvanian* noblemen look upon him as a great natural philosopher, praise his deep knowledge of nature, consult him, and by all those seducing undeserved distinctions hinder him from getting the better instructions by his masters in that part of science. Accordingly his future publications on Natural History will be but too much resembling the former ones. He seems to be sensible that his indigested accounts cannot recommend him to the esteem of true connoisseurs. His chief attention is therefore bent of late on other objects. He intends to make tiles from *asbestus*, paper from different vegetables, and borax from stalactites.

With such proposals he endears himself to some of the *Transsylvanian* noblemen, calls them his generous *Mecænas*'s in proper humility, and is to have from the provincial states an annual pension of 300 florins, in order to realise his projects and to continue his discoveries in Natural History. It is a great advantage for P. *Fridwalzky, in ista loca venisse ubi aliquid sapere videatur.* I desired him to show me the stone from *Gyalupopi* from which he proposes to extract borax. It was a common calcareous stalactites. Then he brought me his *stannum siculum* which he mentions p. 104 of his Natural History. It was a black crystallized blende *(black Jack)* from *Kapnik*, which never will yield any tin. At last he procured me a sight of the crystal, with inclosed gold, which he describes p. 177. As I did not consider this phænomenon to be very extraordinary, I little fancied that even in this he should have been mistaken. But the pretended crystal was common glass, containing within its substance a painted gold garland, such as at a few cruicers expence you might buy by thousands at *Turnau* in *Bohemia*. All this proved to me that the gold grains in grapes, the liquid fluid gold, and many other fantastical singularities which he pretends to have seen, to have examined, or to have heard of, deserve no credit at all.

LETTER

## LETTER. XVI.

*Nagy-Banya, Aug.* 2, 1770.

I SPENT two days in going from *Claufenburg* to this place; and I had no leifure either to fee the iron-works at *Toroczko*, nor the lead-mines of *Runda* near *Biftriz*. I had likewife no time to fpare for the falt-works at *Dees*, becaufe I am to make great hafte, in order to return to *Shemniz* within the fpace of time which I am allowed to be abfent. But I know by very good authority, that thefe falt-mines entirely refemble thofe at *Torda*, and that the falt produced is exported to *Hungary*. All the roads, and all the hills, which I paffed, were covered by a pale yellowifh lime-ftone, containing many marks of broken fhells. In fome parts the micaceous flate, on which it is fuperincumbent, appeared naked above ground. *Nagy Banya* is fituated in a valley, furrounded by a tract of mountains, which runs from the north to the eaft. It is a free and royal mining town, and was in former times, with its dependant mines, the conftant domain and allowance of the queens of *Hungary*. In the ancient records it is

often

often called *Rivulus Dominarum*, on account of a brook which runs along the northern hills, connected with the *Carpathian* mountains. From such records it appears, that its mines have been worked already under King *Lewis* I. in the year 1347. King *Matthias Corvinus* left to the city, in 1468, the mint and the mines for an annual lease of 13000 gold florins. In the *Hungarian* laws, from 1519, two chambers of the royal revenues from mines are mentioned, one at *Kremniz* and the other in *Rivulo Dominarum*. This and the very name of *Nagy Banya*, which is the *great mine*, proves the antiquity of its succesful works. The ancients seem to have been very skilful in smelting and parting their ores. One hundred weight of of clear ore is said to have contained from 79 to 112 ounces auriferous silver. The poorer and mixed ores were in those times ground in common mills; and one hundred weight of the old flags yields scarce a drachm of silver. From the year 1526 these mines decayed by a succeffive variety of accidents, war and rebellions, till in the midst of the last century they were given up entirely. Such was their abandoned state till Baron *de Gersdorf*, one of the most intelligent mining-officers of the imperial states, proposed the renewal of the works in the *Kreuzberg*. Count *Gotlieb Stampfer*, whose character and mineral science you

are

are acquainted with, ventured himself by a small gaping of an old gallery down into the mine, which he found drowned by water. With great danger he shipped in a sort of skiff over the depth to the sides of the vein, and gathering there some rich ores, he greatly encouraged a subscription for draining these works.

To this purpose a deeper gallery was resolved; and the success being unquestionable, and the situation as favourable for smelting furnaces and stamp mills, as the account of the ancients, the proprietors pursue the undertaking with unwearied zeal, and have been these last seven years seriously engaged with their gallery.

It was at first driving for eighty fathoms in grey marlestone; then followed dark grey hardened clay, and at last the metallic rock. This is the only mine at present working at *Nagy-Banya*. Several other fissures however are working by poor adventurers, but at present without success. In the year 1748 a board of surveyorship of mines over several works, formerly belonging to the chamber of *Kashaw*, was established here; and ever since, the neighbouring mines at *Kapnik*, *Felso-Banya*, *Fekete-Banya*, *Lapos-Banya*, and *Mis-Banya*, have been constantly in a more thriving state. I shall examine all these mines, and set out to-morrow for *Kapnik*, which will procure me materials for a longer letter.

LETTER

## LETTER XVII.

*Nagy-Banya, Aug 6, 1770.*

HAVING the choice of all the places belonging to the *Nagy-Banya* inspection, *vicit amor patriæ*. I went to *Kapnik*, a royal mine, in a rough country surrounded with mountains, and situated on the furthest limits of *Transsylvania*, formerly belonging to the *Transsylvanian* chamber, but of late subjected to the direction of *Nagy Banya*. It is four hours journey from this place, and the cliffs, which I met with going there, consist of large naked granite-rocks and argillaceous glimmery slate. *Kapnik* lies in a valley. According to an old tradition the *Transsylvanian* princes are said to have opened the first mine towards the end of the sixteenth century. Its name the *Princes-gallery (Fursten-Stolln)* supports that tradition. The annual produce consisted then of four or 500 marks of silver, containing some gold. But the prevailing ground-waters put a stop to the works, which at last ceased entirely. In 1748 they were resumed, which was occasioned by the *Josephi* mine, as having been sold to the chamber by the impoverished proprietors for the trifling sum of 800 florins. *Kapnik* has at present the advantage over many other mining places, on account of many unattempted mountains, in which of late have been discovered the *Maria Hulf, Barbara* or *Joseph, Josephina, Kapnik* or *Ungarstoln, Erzbach, Theresia, Clemens, Peter Paul,* and *Christopher*'s veins, each working by different companies. All these veins run from north to south, dipping from west to east. The rock of these metallic mountains is a white argillaceous compact stone, resembling our *Saxum metalliferum*, except that it contains some spots of white stone-marrow (*lithomarga*). The rock of the deaf and barren mountains hereabout consists of a blueish trap, striking some fire with steel. In some places both these rocks are covered with micaceous clayish slate. The princes-gallery vein is pursued already a length of 427 fathoms. In the present drift it is narrowed a little by the skirting rocks grown harder. Its common breadth is four or five fathoms. It consists of rosy-coloured feld-spath sprinkled with fallow silver ore. The pure fallow ore is melted separately; and the feld-spath goes to the stamp-mills. In the hading side of this vein runs a lead and blende-vein. Fourteen fathoms deeper are some smaller drifts working; but the rest is drowned in water, which the an-

cients

cients worked out by pumps, and henceforth will be carried off by the deepest gallery, which is to cross and to drain all the veins together. *Petri Pauli vein* produces an auriferous white quartz, sprinkled with fallow silver-ore, and now and then some nests of a pale yellowish pure scarce coherent auriferous calcareous earth. The quartz contains grey plumose and pale yellow antimony, which is likewise found next with the vein in white clay, but then in coarser crystals. I was told that the vein grows richer in gold wherever antimony happens to be found in it. I observed here on the hading side of the vein a crystallisation, whose surface was all over covered with small cubes. You know how curious I am of crystallisations. I endeavoured therefore to separate it from the vein, but found that this whole mass of cubes consisted of a calcareous unpetrified earth. I observed the same phænomenon some weeks before I set out on the old *Anthony of Padua-Stolln at Shemniz*, where in the deepest sole I found a large cube, which at first sight I believed to be a hollow spar-cube, such as I have many in my cabinet. When I touched this crystallisation the smaller cubes, which covered the greater one, were found incoherent, nay liquid, and the greater cube yielded under the pressure of the finger, and the water contained within run off.

off. These facts prove that cryftallifations are continually producing in the humid way; and accordingly the many hollow cryftals, which you have feen in my cabinet at *Shemniz*, have been in the beginning liquid lumps, whose furface cryftallifing in forms convenient to their faline nature, fucceffively hardened and left an empty cryftallifation after the water was evaporated.

The *Maria Hulf vein* is a hardened auriferous clay, fprinkled with pyrites, wafhed in the wafhing-mills.

The other veins confift of a pale reddifh auriferous feld-fpath, commonly fprinkled with fallow-ore. The redder, that is to fay the more irony it is, the more auriferous too. Moft part of thefe fiffures are feparated from the deaf rocks by a foft argillaceous fkirt (*faalband.*) The deepeft gallery at *Kapnik* has been driven already 700 fathoms in length into the above white metallic rock. It was begun by Baron de *Gerfdorf*. The veins croffed by it are large and fine; but their auriferous quality diminifhes in the depth. This gallery is to be lengthened 500 fathoms, and then it will drain the remoteft *Furften Stolln-vein.*

An hour's ride from *Kapnik* in the *Rota mountain*, is a private mine, which I examined likewife. The vein runs between a green indurated

somewhat

somewhat calcareous rock, which is on the hading, and the white metallic rock, which is on the hanging side. It consists of white quartz, mixed with blende and lead-glance. It is auriferous, nay the gold appears often in visible native lumps.

The stamp-mills are in general of the same construction as those in *Lower Hungary*. However, the present upper inspector, Baron *Smidlin*, has built one with six pestels and double gutters. This really does a good deal of work in very short time ; but as the pestels go very briskly, and as more water is spent and to be given in the trough, I do not conceive this construction to be any great advantage, since the impetuosity of the briskly running water does not permit the fine gold-dust to precipitate in the canals, and consequently a good deal is carried away.

They have here three smelting places with eight furnaces, both for the royal and for the private mines; the private reguli to be sold in the royal purchase office, and the whole produce of gold and silver to be coined in the royal mint at *Nagy-Banya*. Their method of smelting is hardly in any thing different from that used at *Shemniz*. Baron *Gersdorf* tried here an experiment, which deserves your notice. He ordered a quantity of the rejected salt-rocks to be fetched from the neighbouring bing-places at *Marmoros*,

and

and when the smelting furnaces were going, some troughs full of it to be thrown upon the works, according to the practice of the assayers, and in order to prevent the silver flying away in the fire. However, the effect did not answer the expectation; rather the silver was lessened one half per hundred weight. But this was owing perhaps either to the neglect or the malice of the common smelters who are natural enemies to any novelty.

## LETTER XVIII.

*Nagy-Banya, Aug.* 22, 1770.

MY long silence is the consequence of an unhappy accident, which was very near putting an end to my life. To examine the common firing at *Felso-Banya*, and the great effects produced by so small an expence of wood, I visited the great mine when the fire was hardly burnt down, and when the mine was still filled with smoke. An accident made me tarry somewhat longer in the shaft, by which the smoke went off. In short I lost my senses, and fifteen hours after I was restored to myself by blisters and other applications. My lips were swoln, my eyes run with blood, and my limbs in general lamed. Without the assistance of a skilful young physician at *Nagy-Banya*, and the great care of the upper administration's inspector, Baron *Gerham*, in whose house I lodge, you would have been deprived of your friend; and the question is still whether he is to be saved. A violent coughing and acute pains in the loins, which alternately put me on the rack, are I fear more than sufficient

to

to deftroy this thinly framed machine. If that fhould be the cafe, then my friend I defire you to have my name at leaft inferted in the Martyrology of Naturalifts.

*Felfo-Banya* is one of the eldeft mining places. Formerly the inhabitants had no trade but mining; and in that refpect they got from King *Lewis* II. a grant of freedom, dated 1523. The public troubles, which put a ftop to all the neighbouring works, did not affect thefe, which continued uninterrupted to the year 1689. In 1690 the emperor *Leopold* bought the works at *Felfo-Banya* for 25,420 florins, granting by public charter eternal freedom from any taxes to the inhabitants: the mines have been ever fince in a thriving ftate. At prefent the *Borkul* and the *great mine* are the richeft. Times immemorial the rock, confifting of grey hornftone *(petrofilex)* is worked by firing, which in the upper foles has caufed tremendous caverns threatning ruin, and not to be faftened by any timber. In thefe ruins, or rather in this *old man*, feveral private adventurers hazard their lives in fearching after remaining ramifications of ftampores. I got from fuch an old cavern a fort of ftalactites, which feems to me extremely problematical. It is very light, refembling in colour a red yellow amber, vitreous and gloffy where broken, refifts acids and the fire without giving any fmell.

The

The *Borkul mine* has its hading and hanging sides of the above hornstone. The vein-rock is zinnopel, different from that in *Lower Hungary* only by less hardness. The vein is one fathom more or less, and contains stamp-ore, which yields but two ounces silver, the mark of silver yielding twenty denarii gold.

The great mine has likewise a vein like that of the pacher-stoln at *Shemniz*. In some places it is six fathoms wide. I went with horror over the ruins and the large rocks detached from the floor, till I reached the third sole, which is less ruinous. On the fourth sole a fissure coming from the hading side, crossing the main vein in an acute angle, strikes it deaf. The stamp zinnopel of this vein contains two ounces of silver, and the mark of silver forty denarii or two ounces and a half of gold. Some times they meet here with ore of sixteen ounces silver. In a hanging fissure of this vein is found fine red, sound or crystallised sulphur, (sandaraca) on white milky crystallised quartz; the same sort of native sulphur on yellow orpiment; white coarse cubic pellucid fluor with inclosed sulphur; grey plumose antimony; the same on quartz-crystals pointed on both ends, and closely sticking together; red and grass green antimony; grey coarse radiated antimony on white pellucid rhomboidal prisms of fluor two or three inches in length,

length, which are perforated by the antimony, and prove that these minerals, however different in themselves, crystallised at the same time. I got likewise from the same vein red fine crystallised magnesia.

I have told you already that the mine is worked by firing. But this firing is different from that practised at *Goslar*, and at *Schlackenwalde* in *Bohemia*. The large ruinous caverns produced by it in those places, as well as on the upper soles at *Felső-Banya*, caused the proprietors to think of a more profitable and less ruinous method of firing. They sink their shaft by chissels, hammers, boring and blasting, to six, nine, twelve, fifteen fathoms. To contrive then a drift, a cut one foot high and deep is made to that side which they want to work. In it they place an iron grate, which they call the *pragel cat*, and it is to be covered with a layer of small pieces of wood, about an inch thick and one foot in length, and then to be set on fire. This softens and loosens the rock, which by iron levers is taken off from the sides and the roof, and facilitates the further firing, so far that no side-cut is requisite. The drift being widened by this method, and the vein over head to be obtained, they begin firing at its sole, covering it by a bed of two or three feet of deaf rock or stampore, and lining the hading and hanging sides with

M a wall

a wall of the same rocks, in order to prevent the flames spreading that way, and causing unnecessary caverns. Then they proceed to set heaps of wood as noticed before. Such a heap consists of about twenty-one or forty-two pieces of wood, three and three in layers a-cross till they reach the roof. Twenty-four heaps requisite for a length of four fathoms; and for these, together with the breaking down the loosened rocks, are paid to the miner twelve one-half cruicer and eight ounces of lamp tallow. By this method only the top of the flame works on the rock. They have an hundred skilful manipulations to direct it to the hanging or hading side, where they want it most. The loosened ores are left on the soles, new soles made of them, new firings set, and thus continued till the whole vein between two soles be broken down. The vein being extremely large, they begin firing in the hanging, going on to the hading side on the same level a-cross the whole vein. The stamp-ores thus loosened are left in the mine as a magazine till they are in want of them, when they are carried under the shafts, and thence to the stamp-works. Thus the ore is procured at a cheap rate, a fathom of a drift by firing costing not above fifteen florins, and requiring only two or three florins value of wood. The solidity and hardness of the rock is an advantage to this

manner

manner of working; cracked and broken mountain or vein rocks would be lefs acted upon by fire.

The vein rock of this large mine is zinnopel, a jafper-like ftone, extremely hard, fcarce to be conquered by the chiffel and hammer, and hard to be blafted. However, I cannot be convinced of the great advantages of this apparently cheap and efficacious firing. Perhaps you will agree with me.

1. Whatever care be taken in lining the fides it is impoffible to prevent the flames from loofening the rock-fkirts of the vein, which, unfupported by timber, of courfe break down, and produce thofe unavoidable and dangerous caverns.

2. The fire volatilifing and diffolving the pyrites and the femi-metallic part of the ore, makes the air of this mine extremely unwholefome, and affects the health of the poor miners, who feldom arrive to any advanced age.

3. The miners cannot, on account of the fire and fmoke, work more than three days in a week.

4. The ores being pyritical, the fire carries too much fulphur away, which of courfe leffens the lech-ftone and the fcorification of the unmetallic parts.

5. It produces likewife an unequal ftamping, fince the fire affects the ore in a very different manner,

manner, calcining some and leaving many more found and unhurt at all. In short, a well directed cross or oblique working would be, according to my opinion, preferable in many respects to this firing.

The great mine is at present sunk 70 fathoms deep. Its vein runs in a metallic rock; and I am inclined to believe that the upper drifts are driven in superincumbent hornstone. The many fragments among the rubbish of the bing-places make me believe so.

The whole mountain of *Felso-Banya* is underdriven or underworked by a deep gallery 454 fathoms length. The ground waters drained by two pole-pump-engines (*stangen-kunste.*)

The royal stamp-mills at *Felso-Banya* are of a good and regular construction. Those of the private companies are but too much resembling to the common bungling stamp-works of the *Transylvanian* gipsies

The smelting places are, one of six furnaces, royal; the other of two furnaces, the property of the city. The gold and silver produce to be delivered into the royal mint.

Companies who work still for hope of a dividend have for a mark of their silver 21 florins 20 cruicers; and for their gold 77 ½ ducats; but those who get a dividend, or for other reasons are deprived

prived of the higher tarif, are paid only 17 florins per mark silver, and 75 ducats per mark gold.

The unhappy accident which befell me here deprived me of the pleasure to examine the other works and the salt-mines at *Marmoros*. What I know of them is as follows:

The salt-mines at *Marmoros* are surrounded with micaceous clayish slate, which continues to the granite-rocks of the *Carpathian* mountains. It contains those fine pellucid octangular alum-formed quartz crystals, which, carried by the rains into the brooks, are gathered there and sold under the name of *Marmaros* stones. They have a considerable hardness and natural polish. In the salt is found pellucid, white, fine striped gypsum.

*Fekete-Banya* is a hamlet belonging to the city of *Nagy-Banya*. In 1645 above two hundred miners are said to have been employed here for gold and silver; but at present it is a deserted place. However, in 1752 the city of *Nagy-Banya* opened a mine, which to this moment is not very successful.

*Lapos-Banya* is a metallic mountain, belonging to Count *Karoli*, and is divided into the *Miz-Banya* and *Sargo-Banya* mines. These last have large stamp-ore veins, running in metallic rock, and richer in gold than in silver. Now and then there is some lead-ore.

*Miz-Banya* has likewise copper. The gold and silver produce to be delivered into the royal mint at *Nagy-Banya*.

In *Olalapos* have been of late discovered some rich veins. I have from them a sample of zinnopel, sprinkled with native gold.

Near *Illoba* is native copper sticking to lead glance. In the *Szamos* river herrings are said to be fished now and then, as has been assured to me by eye witnesses. Near *Nagy-Banya* is a mineral well, frequently used by the inhabitants.

If you knew how much I have suffered to finish this letter, you would consider it as the greatest proof which you might expect of my friendship.

## LETTER XIX.

*Shemniz, Sept.* 5, 1770.

IT is impossible to tell you what I have suffered in my ten days journey from *Nagy-Banya* to this place. I was commonly carried into the coach and out again. Any stone in the road, any jumping of the coach doubled my pains. The dry coughing, which I cannot get rid of, does not permit me any rest. Under such circumstances is was impossible for me to examine the mountains; and commonly the road went over plains in which I had the *Carpathian* mountains to the right, and the *Theissa* river to the left.

*Schmolniz* is on the promontory of the *Carpathian* hills. I should have strayed too much from the road if I had gone there. Besides the recovery of my health, if any to be hoped, did make my haste home necessary.

Near *Tockay*, in the fields and the vineyards, are found pieces of vitreous black and blueish lava, *Pumex vitreus Linnæi*, commonly called hereabout *lux-sapphires*. I did not see hereabout any mountain, which might be supposed a volcano in former times. The hill, which is so famous for producing the noble *Tockay* wine, confifts of argillaceous flate, and does not manifeft any other volcanic marks. This makes me believe that thefe lava pieces have been rolled down and wafhed from the *Carpathian* hills, in which fuch lavas and native fulphur are very common.

A days journey from *Altfohl* begin the hills, which to the right ftretch in the *Luptan* diftrict, and thence join with the *Carpathian* mountains. Here I could not bear the jolting of my carriage; I was forced to ftep out and to creep on very flowly. I obferved vaft quantities of loofe granite pieces. Probably fome of thefe mountains confift of granite; but as far as I might obferve in paffing by they are in general covered with argillaceous flate. Some lead-mines are working here; and as they told me fome copper-mines too.

When I reached *Shemniz* the account of my accident arrived likewife; and as report heightens matters, I was faid to have not furvived it. My arrival was therefore a welcome confolation to my family. Among a great many letters, yours were opened firft: I rejoice on your happy return into your country. I have partly fatisfied your demands by the accounts of what I have feen and obferved; but I do not know whether I fhall be able to give you a compleat theory of the nature
of

of the *Hungarian* mountains. However, I will endeavour my best in one of my next, and for the present I shall entertain you with the works at *Smolniz*, as described by me some years ago, when I wanted principles and experience for such a task, and when I little thought that such observations might prove an advantage to Natural History.

*Smolniz* is a celebrated mining place, belonging to the royal domain, situated in the district of *Zips* on the foot of the *Carpathian* hills. It has very noble copper-works, known in the times of Count *Zapolia* and *Bathori* under the emperor *Ferdinand* III. The *Zips* county and *Smolniz* devolved to the Counts *Csaky*, who in general lett the mines and the copper trade to foreign tenants. About the year 1671 *Francis* and *Stephen*, Counts *Csacky*, divided their county and the *Smolniz* mines in two equal parts; but *Francis* having engaged in count *Tokely*'s rebellion, his part was forfeited to the crown. But the imperial chamber had no proper idea of this mine, and accordingly left this part to several private tenants, and in 1684 to the comproprietor count *Stephen Csaky*, at a yearly rent of 4000 florins. Three years after the chamber worked this mine at a common expence with count *Stephen*, and its share of the annual dividend was 14831 florins. This opened the

the eyes to the chamber, and made it think of the acquisition of the other part, which was obtained by an exchange in the year 1690, when the confiscated part of the county was returned to Count *Stephen*, and he prevailed upon to renounce his part of the mine, and the hamlets *Smolniz* and *Stoofs*, into the possession of the imperial chamber. Since this transaction the chief mines at *Smolniz* were royal estates, under the direction of the chamber at *Kashaw*; however, by want of proper principles, they could not be brought to any remarkable dividend. Therefore, in 1748, a formal general direction was established here, and the works were entrusted to a man of experience. The mines belonging under the general direction of *Smolniz*, are *Smolniz*, *Stoofs*, *Swædler*, *Einsiedler*, *Gollniz*, *Krumbach*, *Borathshod* or *Wagendrussel*, in the county of *Zips*, *Under* and *Upper Mezenseiffen* and *Jossaw*, in the *Abavira* district; *Tobshaw*, *Rosenaw*, in the *Gomorha* district; *Iglo* or *Newdorf* and *Wallendorf* under *Polish* jurisdiction.

The mountains at *Smolniz* consist of a blueish glimmery argillaceous slate, in which the three veins at *Smolniz*, called the middle, the exterior lying, and the exterior hanging vein are running in hour six in the morning, dipping to the horizon in nearly seventy-five degrees. Hence it appears

pears that they are parallel. They are about twenty fathoms diftant from each other; in their bendings they are ftill nearer. The fiffures between thefe veins are of no remarkable metallic value. The veins are fubject to leaping, running fometimes deaf a great length, and are affected in their run even by the leaft change in the fituation or nature of the fkirting rocks. Thefe changes in the fituation of the rocks are owing to fmall fiffures, which cut the run of the vein in a different hour or angle, and are called here *kleins*. By long obfervation of thefe crofs-fiffures, and their effect on the main veins, the following rules have been laid down:

If the crofs-fiffure be coming from the eaft, it drives the vein into the hading; if from the weft, it pufhes the vein into the hanging fide, where accordingly it is to be found.

Fiffures in hour nine or twenty-one, and dipping to the eaft or the north, are called here *irregular* and *refractory* ones; *regular* fiffures in the contrary are thofe that keep the above run and dipping of the main veins.

Regular fiffures, falling in with the vein, pufh it into the hading; but irregular ones into the hanging fide.

Though thefe crofs-fiffures or *kleins* very often interrupt the run of the vein, they are however
requifite

requsite to enable them, as generally they are deaf where these crofs-veins are wanting.

The middle vein is the richeft. The fecond in rank and value is the hanging; and in refpect to thofe the lying vein, as running beyond the mountain in the valley, is poor.

The rock of thefe veins is a dark grey clay, commonly mixed with quartz, but very feldom with fparr. Generally the quartz begins and ends the ores.

The argillaceous flate between the veins contains frequently confiderable nefts or lumps of pyrites. I have met with fuch a lump, improperly called here a pyrite-ftock, in a depth of fifty-feven fathoms, between the middle and the lying vein. They are rich of fulphur, and contain two pounds of copper in a hundred weight.

The ores are generally yellow copper-pyrites, either fuperficially variegated, or yellow and found or fprinkled in glimmery dark-grey flate. Befides thefe ores, whofe richeft fpecies contains about twelve pounds of copper, the mines at *Smolniz* produce annually about an hundred thoufand pounds of cemented or precipitated copper.

The water which foaks through the fiffures and veins is impregnated with copper-particles, diffolved by the vitriolic acid. To impregnate it ftill more, it is led into fome old fhafts, thence
raifed

raised by pump-works, conducted through several bing-places, and then poured in canals, which are dug near the shafts and filled with old iron. The vitriolic acid nearer related with the iron, is attracted by it, and accordingly precipitates the copper under the form of a soft powder. If the cement waters be strong, this powder or sediment is every third day separated from the iron, to prevent its being incrustated with copper, which would hinder its further dissolution. It is observed that the copper precipitation succeeds sooner and better in those canals, where the vitriolic water falls with some impetuosity on the iron. Every month the copper sediments are gathered in the canals, the iron cleared and put again in the water, till it be entirely dissolved.

Another equally profitable work is the picking up and washing of the bing-rubbish, consisting either of the neglected ores of the ancients, or of those, which having been thought unfit for separation in the mines, are carried to the bing-places. These places are examined, the found ores separated from the deaf rocks, and the copper sprinkled slate pieces washed and sieved as at *Shemniz*. The remaining unshining pieces go to the mills. This business is carried on by children, servant maids, and old maimed miners, and produces every year about 60,000 pounds of copper.
The

The stamp-mills are of the same construction as those in *Lower Hungary*.

Those ores that are too sulphurous are by roasting separated from the sulphur. Several sulphur furnaces are built for that purpose. They are from three to six fathoms in length; from one to two fathoms deep, and two fathoms high. Commonly they have thirteen windows; and each window several openings, by which the smelted sulphur runs out. The bottom of these furnaces is according to their bigness filled with three or four fathoms of wood; these wood layers are covered with some carts of charcoal; then comes a layer of found sulphurous ore one foot thick, and this is covered with alternating wash-ore-beds till the furnace be filled. The wood below is set on fire by a wood canal vertically set a-cross the ore-beds. Such a sulphur-furnace contains about 500,000 pounds of ore, which continues burning twelve or fourteen months. If the sulphur ceases running from the openings, canals are dug in the upper covering of the furnace and paved with flat stones. These canals, and wooden conductors laid in them, are led to a stone reservoir; and as the rising sulphur steam is catched in the conductors, and cooled in the reservoir, a greater quantity of sulphur is got by this method than by the former. The depuration and sublimation of it

into

into *flores sulphuris* are known practices. The annual produce or saving of sulphur is about 200,000 pounds.

The ustulated pyrites serve here likewise in a vitriol manufactory. When still warm they are put in tubs with water, elixiviated into a brine, and this boiled and evaporated in lead pans, which produces a blue-greenish vitriol, but the want of sale makes this manufactory less considerable than it might be under more favourable circumstances.

The common copper-ores require ten different roastings, one raw, and one black-smelting; after which they are refined with some lead, as not bearing the expence to part the little silver which they may contain.

The richer silver-mixed copper-ores are parted as in *Lower Hungary*, and yield at a medium from twelve to 1400 marks of silver every year.

*Stoofs*, in the district of *Zips*, is a dependant place of *Smolniz*, and furnishes yearly about 500,000 pounds of iron, which for a great part is consumed by the cementation works at *Smolniz*. The iron-veins are here running in slate; the ores consist of brown or red iron-ocher, in a greater depth hardened into an iron-coloured sound-ore, *Hæmatites cærulescens*. It contains now and then some nests of yellow copper-ore.

*Swadler*,

*Swadler*, a market-place, is surrounded with large forests, and a great advantage to the royal chamber. The furnaces in the midst of these forests smelt and refine every year about 200,000. pounds of rose-copper. The company of private adventurers work here in rich copper veins, inclosed in glimmery argillaceous slate, which every year produce above 400,000 pounds of fine copper.

*Einsidel*, a place belonging to Count *Csaky*, in the district of *Zips*, has rich copper-veins, which in former times produced considerable dividends, but for want of wood are but slowly working at present. However, they yield and sell every year to the royal purchasing office about 200,000 pounds of copper.

*Golniz*, in the same district, likewise belonging to Count *Csaky*, is in time anterior to *Smolniz*. It has two rich copper veins, let to several companies. The royal chamber has large shares in them, and some considerable independant mines.

The veins run to the east between horn-slate for a length of 900 fathoms. They dip into a considerable depth. The vein rock is grey quartz mixed now and then with spar. They yield yellow copper-pyrites, and grey copper ore, called hereabouts white ore, which contain fifteen pounds of copper per hundred weight, and from

five

five to twelve ounces of silver. The whole annual produce of this place is about 600,000 pounds of copper.

*Krumbach,* in the same district, belonging to the same master, has iron veins in argillaceous slate-rock. In sundry places of this manor are rich copper veins, which annually produce about 200,000 pounds of copper.

*Beratshod* or *Wagendruſſel* and its dependencies produce about 300,000 pounds.

*Under-Mezen-Seiffen, Joſſaw* and *Upper-Mezen-Seiffen,* in the *Abavira* county, are three mining places belonging to the convent of *Premonstrate-friars* at *Joſſaw.* This county is contiguous to the *Carpathian* mountains; the ground, consisting of argillaceous slate, rather working for iron than for copper-veins.

*Topshaw,* in the county of *Gomor,* on the river of *Gollniz,* has two capital copper-veins in argillaceous slate, worked by private companies. The different mines, belonging to this bailiwick, deliver annually about 100,000 pounds of copper to the royal office at *Smolniz.*

*Rosenaw* belongs to the Archbishop of *Gran,* and is situated in the county of *Gomor.* In the territory of this place are copper, gold, and antimony veins. The large old bing-places at *Zingobanya,* near *Rosenaw,* speak rich old copper-mines.

Some years ago they have been taken up again, and yielded good copper-ores, containing some silver; but the adventurers being unable to afford the expences of pump-engines the works are dropt. The gold-veins discovered some years ago are dropt likewise. The four antimonial-veins run in hornstone-slate. The ore is commonly found granulated and grey antimony; it scarce ever appears in crystalline or plumose forms. Near *Krafznaborka,* in the county of *Gomor,* is a rich quickfilver-mine, which yields fine cinnabar-ore.

*Iglo* or *Newdorf,* in the county of *Zips,* is one of the thirteen towns which the Emperor *Sigifmund* pawned to the crown of *Poland.* It has rich copper-works.

*Wallendorf* likewise pawned to the *Polifh.* The mines belonging to this place work still in hopes of a dividend.

The buying of the copper produced in all these places is a royal prerogative, according to which the different private companies are obliged to sell their copper in the royal office at *Smolniz* at different prices, settled according to the different circumstances and the different goodness of the copper. These prices vary from twenty-nine to thirty-one florins; but the *Iglo* and *Newdorf* copper sells thirty-two florins thirty cruicers. Three months after delivery the companies are paid in ready

ready money; which regular and constant sale has brought these *Upper Hungarian* copper-mines into their present flourishing state. The produce of the private companies consists every year in 1,400,000; but that of the royal mines of 700,000; in all of 2,100,000 pounds of copper.

## LETTER. XX.

*Shemniz, Sept.* 7, 1770.

YOU will not expect a compleat history of the *Lower Hungarian* mines, their origin, works, engines, furnaces, œconomy and produce; that would be a work of some years and of some volumes. Besides, you have been here and seen and examined yourself. You have read Mr. *Severini*'s treatise of the ancient inhabitants and the origin of the mines at *Shemniz*; and you may expect an account of the *Lower Hungarian* engines from Mr. *Poda*, who is resolved to give it to the press. Perhaps a description of our subterraneous and metallurgical-works might appear likewise, if the imperial order be brought into execution. It is, that the professors of our miner-academy are obliged to penn down proper hand books for the disciples, which are intended to be published at her majesty's expence. I want only to remind you of the nature of our rocks, of the rule of our veins, and of some observations thence arising, in order to support a short theory of all the *Hungarian* mountains, which

which to comply with you, I have resolved to sketch out.

The promontories of the hills, in which the noble veins at *Shemniz* are running, raise on the borders of the *Gran* river, where they consist of slate, which afterwards unites with a harder, argillaceous grey rock, mixed either with sherl, or quartz, or calcareous spar-grains.

This rock, which hitherto I constantly have called metallic rock, is the common mountain-rock, in which the veins at *Shemniz* and *Kremniz* are running. They unite with the *Carpathian* mountains.

In the valleys behind *Shemniz*, near the glass manufactory, and in several other places of the *Lower Hungarian* mines, hills of grey limestone are accumulated on the sloping, nay even on the summits of this argillaceous rock.

The metallic rock near *Shemniz* contains three capital veins, parallel in their run with the direction of the river *Gran*, nay even with the bendings of its channel, as clearly appears by the mineralogical map of the *Shemniz* veins, which has been published by Mr. *Zipser*.

The largest is the *Spitaler* vein. It runs from north to south between twelve and four, and dips from west to east from thirty to seventy degrees. In the remotest northern field this vein, belonging to the *Nicolai* mine, was crossed by a drift, and found deaf in the depth. Here it consists of grey loose and soft clay, mixed with spar. But the old rubbish seems to indicate that there were likewise rich and metallic veins in the upper level towards the day. Somewhat more to the south, in the fields of *Michels* and *Pacher-Stolln*, it furnishes good stamp-ores. Here the vein consists of quartz, leadglance, and zinnopel; and the washed metallic powder contains a good deal of gold. In the field of the *Pacher Stolln* it is larger than any where else. On the twentieth drift or parallel for example, at a depth of 127 fathoms from the *Elizabeth* shaft, it is fourteen fathoms wide, and the deaf rocks included eighteen. This remarkable thickness of the vein has caused the cross-works to be introduced. The company of the *Three Kings* and *Pacherstolln* has here in the depth of the thirteenth level or drift, that is to say, in the depth of eighty-seven fathoms very rich ores, which in a greater depth change in zinnopel stamp-ores.

Further to the south, near the limits of the *Windshaft* field, an argillaceous white vein unites, and thence constantly runs with it in the hanging. The vein is from that place found to contain silver. The white clay of this hanging fissure offers now and then nodules of spar and quartz pieces.

Containing

Containing four or five ounces of silver per hundred weight; they are carefully gathered. At last the form of the vein is entirely altered in the *royal wind-shafts* field; and it consists of broken quartz, which, if mixed with spar, yields richer ores of dissolved pyrites and of an irony ocher, which seems to contribute to its increasing auriferous quality.

This irony ocher decreases in the same proportion as the vein is advancing to the south. The remotest field to the south, distant from the northern about 3000 fathoms, is entirely deaf.

The *Johns* vein runs in a distance of 1000 fathoms in the hanging of the main *Spitaler* vein. According to Mr. *Zipser*'s map, this vein seems to be the same which I mentioned before, as uniting or issuing from the chief vein. The vein-rocks or substance of this fissure is white clay, containing sometimes ore, sometimes metallic quartz, and sometimes sprinkled metal flakes. In its midst occur now and then separated and deaf zinnopel fragments, or dissolved zinnopel. To the south, on the fourth and fifth drift or level, its substance turns harder and richer; however, it has constantly in the hanging an argillaceous skirt or separation of two or three inches thick, and in the hading a skirt of zinnopel about one foot thickness.

The mines working on the *Spitaler* vein are: the *Windshaft*, *Pacher-Stolln*, *Three Kings-Stolln*, the *Glanzenburg*, *Michels-Stolln*, and the remotest northern *Dillner* deep gallery-works. I am to take notice here of a singular curiosity, which is that in a drift driven on this vein, and in a perpendicular depth of eighty-nine fathoms from the *Elizabeth* shaft, I have found included in the sound zinnopel a species of petrified porpites. * I am possessed of a fragment of zinnopel with several impressions of this marine body, and a porpites belonging to one of these impressions; both offered to me in a bing-place of the *Spitaler* vein, among the rubbish and stamp-ores, which then were carried out. The day after I went myself into the mine, asking the workmen and officers whether this sort of impressions and stones had offered to them in working out stamp-ores? They answered me in the affirmative, but unconcerned at their singularity had constantly rejected them under the stamp-ores. However, my endeavours to find some other specimens for my mineralogical friends have been to no purpose.

* It is properly a petrified *Madrepora simplex, subtus plana, annulis concentricis, supra convexa, umbilico impresso, lamellis approximatis* in *superficie granulosis*. The same species, tho' hitherto undescribed, is common near the salt-works at *Gemunden* in *Upper-Austria*.

Now,

Now, my dearest friend, tell me how these petrifactions may have been brought into a vein of a mountain, which cannot be ranked among the accidental mountains, and whose very rocks prove it to be of the primitive or original kind. If there were any higher superincumbent calcareous hills, this phenomenon might prove less singular. But our calcareous hills are on the sloping of the mountain towards the glass manufactories, and these petrifactions cannot possibly be carried thence to this high elevation, as none of that kind are to be found in the calcareous and lower strata. However, a single circumstance seems to give some explication. You remember a hill to the north of *Shemniz*, which of late is decorated as mount *Calvaria*. It consists of micaceous clay shistus, mixed with detached pieces of red jasper, which greatly resembles the deaf zinnopel. Petrified turbinites and chamites have been very often found about this hill; and samples of these petrifactions are in my cabinet. Might not, when this accidental hill was produced, and the whole country was still under seawater, some of these crustaceous animals have been carried by water in the still gaping fissure of this large vein? and by that accident have been brought into the mass of zinnopel? § The detached jasper

pieces

§ If these petrifactions were of the same kind as those in the before-mentioned hill, which is not, or the species of turbinites pieces, mentioned before, seem to have been brought by similar accidents in the *Johns'* fissure in the hanging of the vein.

The second or the *Beaver Stollner* capital vein runs in a distance of 100 and 150 fathoms, in the hading of the *Spitaler* vein, parallel with it both in its run and in its dipping. The vein-rock is quartz mixed with pale yellow and reddish spar, zinnopel, and richer ores. In the hanging towards the *Spitaler* vein it contains zinnopel and lead stamp-ores; but in the hading is a skirt of clay, from one to four foot thick, with nodules of lead-ore, which yields from two to five ounces of silver. This vein has not been worked, nor found metallic so far to the north as the *Spitaler* vein. Near the *Amelia* shaft the *Daniel* fissure, straying from the *Theresia* vein, unites with the *Beaver Stollner* vein, and makes it richer. More to the south in the *Christina* field this vein has undoubtedly yielded the greater quantity of ore. Hereabout several regular and irregular fissures, such as the *Althandle-*

---

turbinites and chamites, which are common in that hill, were likewise so in the zinnopel vein, which Baron *Born* would certainly have taken notice of, then the slate-hill and vein-rock of the *Spitaler* vein might be considered as produced in the same time. However, the vein rock may have been produced by similar circumstances, but in times anterior to the origin of that slate-hill. (Transl.)

*Roshka,*

*Roshka*, and some other hanging and hading ones, sometimes unite and sometimes separate from it; especially the *Wolf-gang* vein parts from it and runs towards the *Spitaler* vein. An hundred fathoms more to the south, in the *Siegelsberger* field, this vein has its greatest width, which is from four to five fathoms. However, it is crossed by two small fissures. They worked the first by galleries, and found on the ninth drift a vertical vein, till then unknown, which has ever since yielded the richest ores. The three hanging fissures, straying from and running parallel with the main vein, yield likewise great quantities of rich ore. Beyond the *Konigsegger* shaft to the south the vein is deaf at present. But the southern part of the mountain is reserved for the chamber, in order to examine the vein and to lengthen the works.

The *Theresia* vein runs in a distance of 150 fathoms further in the hading, in a direction parallel to that of the two before described. On the highest part of the mountain it bassets out, and is less examined than any other. To the north it dips from east to west, that is to say, in a contrary and irregular direction; more to the south it turns perpendicular; and further beyond the *Theresia* shaft it dips parallel to the other capital veins at *Shemniz*, that is to say, from west

to

to eaft. The vein-rock is lead-ore, mixed with zinnopel; however, in fome fiffures, efpecially *S. Daniel*, it has yielded richer ores. Its foundnefs is the caufe that the works on this vein are not funk above 150 fathoms, that is to fay, to the fole of the *Trinity* gallery. The zinnopel veins being richer of gold make amends for their being poor of filver. That is the cafe in this and the *Spitaler* vein. On the eleventh level, that is, in a depth of 116 fathoms, they have found in the midft of the found mountain rock, when cutting a crofs door, or an oblique drift, a great number of bullets from one to two inches thick, confifting likewife of metallic rock, and being either loofe or fticking to the other rocks. I am at a lofs to guefs how thefe globular ftones came into the mountain, or why they came to be produced in this form rather than in that of the other compreffed rocks. On the *Klingerftolln*, belonging to the *Therefia* fhafts field, I obferved the remains and marks of old works, done before the invention of gun-powder, which the ancients called *pocket works* (*tafhen arbeit*) and produced by ftrong and dry wood wedges, forced in the hanging fide of the doors or galleries. Thefe wedges being wetted afterwards broke the rock by an effect of their expanfion.

I remember that in the description of the *Spitaler* vein I mentioned deaf wedges, which frequently offer in the greater veins at *Shemniz*. They consist of an argillaceous grey rock, differing from the metallic rock in a single circumstance, that instead of mica it contains spots of white lithomarga.

The width and thickness of these veins has probably occasioned the great quantity of crystallisations which are so common in these mines. Being filled with metalliferous substances there might have remained, or by their drying have been produced those holes in which the cryftallisations are deposited. They are generally filled, or at least incrustated within by calcareous, selenitical, or quartz cryftallisations. Supposing that their regular forms are owing to salts, as commonly is supposed, the variety of metals and semi-metals of the large veins at *Shemniz* seems to have naturally produced that great variety of fine cryftallisations which you have seen in my cabinet.

Thus I have communicated to you my observations on the chief veins at *Shemniz*. However, I add that they are worked with success to the greatest depth, since the auriferous zinnopel continues in vast quantities. Though this depth be about 200 fathoms, it is of no great importance in respect to the inner structure of the globe, if

we

we consider that, according, to Mr. *Pedas* barometrical observations, made under ground, the seventh sole in *Sargozi* mine being 158 fathoms below the *Charles* shaft, and 286 fathoms below the *Therefia* shaft, is still much above the level of the city of *Vienna*. Accordingly our subterraneous geographers have no reason to be proud of their discoveries of the inner structure of the earth; even our deepest shafts have at most but scratched its exterior surface, and *Ovid's*

*Itum est in viscera terræ* will be a poetical extravagance, till *Maupertuis's* unphilosophical proposal of a shaft sunk through the centre of the earth to the *Antipodes* be brought into execution

All the royal mines at *Shemniz* are drained and underworked by the Emperor *Franciscus's* gallery. Its door or entrance is five *English* miles from *Shemniz*, in the *Hodrizer* ground. It was begun in 1748, and happily finished in 1765. Continually driven in sound and hard metallic rock, and in a considerable height and width, it is to be wondered at even by connoisseurs how this immense and difficult work could be atchieved in so short a time. It is lengthened still further.

The private mines, working in other metallic veins, and belonging to the metallic court at

*Shemniz,*

*Shemniz*, are to the west or to the north in the *Ross ground* near *Eisenbach*, or to the south in the *Hodriz*. The *New Hope-Stolln*, the *Hofer Erb-Stolln, Windshleuten,* and the *Old Antony de Padua Stolln*, are the most considerable mines in the *Ross ground*.

The mountains consist here generally of metallic rock; the veins, like those at *Shemniz*, running from north to south.

In the southern valley is *Old All Saints Stolln*, a royal mine, which in former times yielded very rich ores, whose remains are still working. The co-incidence of some large veins has produced here vast caverns, supported by the sparings of deaf rocks. Here I found in a pit the year 777 cut in the rock. Might I hence draw a consequence for the high antiquity of this mine? The rocks are here likewise our argillaceous grey rock, with this difference however, that it is as in all the other *Hodrizer* mines mixed *Lithomarga*.

Next to this mine follows *Finster Ort* and *Brenner-Stolln*, whose vein-rock consists of large auriferous and lamellated quartz, extremely light, very often as it were cut with a knife, but commonly in a state of corrosion, and containing in its holes rich auriferous red and brittle glass silver ore. The *New Antoni de Padua Stolln* has a similar quartz vein.

More

More to the west our metallic rock is covered with argillaceous slate, crossed by iron and lead-veins. The same appears in the *Ross ground* valley, where the slate begins near the *Eisenbach* bath, and yields iron-ore, and now and then some fine load-stones.

I pass silent over many smaller private mines to speak now of *Belo-Banya* or *Dulln*, a mining place to the north of *Shemniz*. The mountains the same as at *Shemniz*, but an hour's way distant. The mines do not seem to have been of great importance in former times; and nothing is left but the names of some mines and the still working deeper *Dullner Erb Stolln*. The *Dullner Nicolai* shaft is sunk on the *Spitaler* vein, and belongs for the greater part to the royal chamber. I saw here on the *Seven Women* vein, in the northern field drift, several old blasting-holes, one marked with the year 1637. *Rossler* § relates, that in 1627 the blasting of the mines was brought from *Hungary* and introduced in the *German* mines. But *Bayer* says, that in 1613 it was invented by *Martin Freygold* at *Freiberg*, an assertion which is repeated of late in the account of mines and their workings, published by the miner-academy at *Freiberg*.

§ Berg-Spiegel.

The *Maria Hulf* company works on a vein of auriferous pyrites. It runs from north to south.

To the south, on the other side of the *Reichawer* water-reservoir, is a private mine, called the *Moderstolln*, on a vein of auriferous quartz. It has given good dividends.

Further to the south is *Baka-Banya* or *Bugganz*, a mining town. The mountains, like those at *Shemniz*; some veins running to the east. Sundry companies have of late taken up different old mines, in auriferous quartz and spar veins. The *Ladislai* company produces already a good deal of gold, which often appears in a granulated form.

Behind *Bugganz* the sloping of the mountain is argillaceous slate, dipping under the plain which by *Tyrnaw* continues to *Presburg*. On the northern side of *Presburg* appear the roots of the *Carpathian* mountains, which near *Modern* are working for lead veins, mixed with asbest and running in horn-slate.

I have noticed already in my last letter the *Schistous* mountains to the east of *Shemniz*. To the north, on the sloping of the mountains, are calcareous hills of white fine-grained limestone. They extend to the *Glass-huttner* bath, which consists of a hot well, used for many infirmities. The water deposits tophus mixed with iron ocher, and the canals in which it is conveyed to the bathing-house are incrustated by it. Many tophaceous hills of the same kind are seen hereabout, and probably have been accumulated on the surface before the water was collected into canals.

Between *Creuz* and *Lehotka* is a fine cultivated plain. Marks of coals have been found here, and mentioned already by *P. Kircher* in his *Mundus Subterraneus*. Near *Lehotka* and the highway I found white shivery hornstone, petrosilex, resembling chalcedony and containing some petrifactions of plants or corals. Probably these detached loose stones have been carried hither from *Leskowiz*, beyond *Cremniz*. A rivulet, which flows here, comes that way; and found beds of such milk-white hornstone are found near that place, besides a great number of detached jaspers and achates, which are very common in the fields thereabout, and near *Deutsh-Littaw*.

The mountains at *Kremniz* are our common metallic-rock. The works are on a large and rich gold-vein and on some of its ramifications. The rock is white solid quartz, mixed with fine auriferous red and white silver-ore, and with auriferous pyrites. This pyrites, properly stampt and washed, contains from two to three drachms of gold per hundred. The vein runs from south to north, and is auriferous above a length of 3000 fathoms

fathoms. It has been searched already to a depth of 150 fathoms, and is constantly found auriferous.

The same vein is, besides the royal mines, worked by the *Rothish* company and the citizens at *Kremniz*. Fine striated grey antimony found in the king's shaft. The whole vein being metallic, a great number of stamp and wash-mills are established here, and to great advantage directed by your countryman Baron *Watram*.

Further to the north, near *Tshavoja*, some lead-veins are working in blue micaceous clayish slate, probably superincumbent on metallic rock.

To the east *Kremniz* is separated from *Newsol* by a steep mountain, consisting of metallic rock and covered with slate. At its top, called the *Skalka*, red native sulphur has been found in a grey superincumbent sand-bed. On the other side of the hill is *Tajova*, known by the royal copper- and parting-furnaces. Near this place crystalline orpiment, *Auripigmentum crystallisatum, crystallis polyedris*, is dug from a blue clay vein in slate. The hills thence to *Newsol*, an hour's way distant, are calcareous.

This city is in a pleasant plain on the *Gran*.

To the north, near the village *Jacub*, rises the mountain *Baran*, consisting of argillaceous slate, accumulated on limestone and containing some copper-fissures. Nearer to *Herrn Grund* appears again the grey micaceous clay shistus, in which the mines are working. There are three chief veins, running from north to south, and dipping from forty to fifty degrees from east to west. The vein in the hading is called the *Copper-vein*; the second to the hanging the *Herrngrunder-vein*; and the third, more in the hanging, goes under the name of the *Pipe-Stolln-vein*; to which may be added a fourth, further in the hanging, and called the *Rat-ground-vein*. All these veins are cut off to the south by an oblique crossing vein of red irony argillaceous slate, which is many fathoms thick. Beyond this red vein they have begun a gallery, unsuccessfully driven already 278 fathoms length in black limestone, in hopes to reach again the copper-veins, if they should be found continuing on the other side of this red vein. The vein is common shivery clay, different from the mountain by a small mixture of mica. Often the ore is sticking to quartz and gypsum. The ore commonly copper-pyrites, called here *gilf-ore*, containing no silver, but from eight to ten pounds of copper in a hundred. There is likewise grey copper pyrites, containing sixteen or seventeen pounds of copper, and from three to ten ounces of silver. Accidentally they meet sometimes with fine samples of white ore, *Cronstedt*. §. 199, crystalline grey copper-glass, *Cuprum vitratum crystallisatum crystallis decæ-*

*dris*

*dris & planis tetraëdris*, noble famples of verdegreafe and copper azur, malachites, and white hair or capillaire vitriol, *Halotrichum Scopoli*, fweating from or fticking to the fides of the works. This vitriol, which in undeterminate blue and green cryftals is likewife produced on the timber, contributes greatly to its prefervation, fince times immemorial no repair of timber has been wanting. The copper-ores at *Herrngrund* are auriferous. For this reafon they feparate the gold-duft in the wafh-mills. It would be impoffible to feparate it to advantage by fmelting and parting. The whole mine is divided into fix fields, three to the fouth and three to the north. In the northern fields the *Kugler*, the *Pipe Stolln*, and part of the *Herrngrund* veins are working; in the fouthern they work only on the *Herrngrunder* vein. This vein is twelve fathoms wide. However, the different direction of the rocks, which it croffes, affect it fenfibly, either interrupting its run or forcing it into an other line of the compafs. The fame happens to the other veins. It has been conftantly working thefe laft five hundred years to a depth of 150 fathoms. The *Pipe-Stollner* vein is fome fathoms thick; but the *Kugler* vein is only four feet. The cementing water is conducted by floping wood-canals, and many angulated windings in large wood refervoirs. In the corners of thefe canals, and in the refervoirs, they put old iron, which precipitates the copper so successfully, that the sediments contain near seventy pounds of copper per hundred. The annual produce of this cementation does not however amount to above 5000 pounds. There is another profitable establishment in the *Herrngrund* copper-works, a manufactory of verdegrease or mountain-green, *Viride montanum nativum.* To this purpose the mine-waters are conducted on old bing-places, which impregnate them with vitriol by the solution of the copper flakes sprinkled in the deaf rocks. Thence they are led in several wood-reservoirs, where running against obliquely erected planks the green precipitates as a sediment. An hundred weight of this green sells for 100 florins, and is to that rate delivered to the mineral ware- and sale-house at *Vienna.*

The negligence of the ancients in separating the ores has likewise in this place caused many bing-washings. Tho' there is a great number of them, they do not produce annually but about 300,000 pounds of clear copper.

Travelling from this place to *Moditska* towards *Liptaw,* you see on both sides of the way a chain of calcareous stalactite-like hills, above thirty fathoms high. Probably this stalactite has been carried hither by rain-water, from the limestone, superincumbent on the *Carpathian* mountains.

I am

I am however at a loss how to explain the figure of this cryſtalline limeſtone, which exactly reſembles that of ſtalactites, and appears often in globular and columnar forms of one or two feet thick.

Beyond *Moditſka* is a lead vein in limeſtone, worked without ſucceſs.

The iron works near *Rhoniz* and *Thaiſolz* belong likewiſe to the chamber at *Newſol*. At *Rhoniz* the iron-veins run in ſlate. The richeſt is on the *Sirk*, and produces ſome iron ſpar. I have not ſeen *Thaiſolz*. The hills about *Newſol* are calcareous and ſuperincumbent on argillaceous ſlate. But nearer to *Shemniz* the common rock is again metallic rock.

On the midſt of the way to *Shemniz*, where the highway to *Kremniz* and *Newſol* ſeparates, I obſerved near the bridge over the *Gran* ſome rocks of breccia, conſiſting of argillaceous and micaceous blunted ſtones, and reddiſh granite pebbles ferruminated by lime. Probably they have been carried hither from the *Carpathian* mountains and depoſited by the river in the before-mentioned rocks.

Near *Poinik*, an iron work under the chamber of *Newſol*, the iron vein runs in ſlate, and produces theſe fine iron ores, incruſtated with blueiſh diſtilled chalcedony, which you have taken notice of when you were in *Hungary*.

*Konigsberg (Ui-Banya)* in the *Borsher* district, is in rank the seventh of the free *Hungarian* mining towns. It is some miles to the north-west from *Shemniz*. The valley wherein it is situated consists on one side towards *Shemniz* of metallic-rock, and on the other to the north of granite-hills, running hither from the *Carpathian* mountains. In the royal mine, at present working again, the vein runs between the red granite its hading, and between the metallic rock its hanging side. They call this granite mill-stone; since its feld-spath particles being dissolved, and having left many holes, make it a good mill-stone. To this purpose it is exported to many parts of *Hungary*. The vein is grey quartz, mixed with auriferous pyrites. This place is remarkable on account of the first steam or fire-engine established in the *Lower-Hungarian* mines, built here 1721 by *Isaac Porter*, an *English* engineer in the imperial service. Its object was the draining of the *Althandler* shaft; but, the works being given up nine years after, the engine has disappeared of course.

These are the notices which I have been able to take in respect of the *Lower Hungarian* mountains, veins and mines. If I should recover my health next summer, I intend, in Mr. *Poda's* and *Scopoli's* company, to make a trip in the *Carpathian* mountains; *Scopoli* to gather plants and

and infects; *Poda* to make physical and mathematical experiments; and I for my part to have an eye to the nature of the mountains and the fossils.

The descriptions of the metallic-mountains in the district of *Liptaw*, given by *Bel*, in *his Notitia Hungariæ*, and several curiosities which some students have brought me from the *Carpathian* mountains, prove to conviction that such an excursion will be an advantage for Natural History. Were this part of science better cultivated in *Hungary*, this kingdom might, I am confident, furnish more remarkable observations, than perhaps any other in the World. But, alas! scarce the name of that science is known hereabout, and I fear we may for a long while repeat with old *Seneca* to the lazy *Hungarians*: " *Sicut barbari ple-*
" *rumque inclusi & ignari machinarum segnes la-*
" *bores obsidentium spectant,* nec quo *illa pertineant,*
" *qua ex lonquinquo struuntur, intellignnt, idem vobis*
" *evenit. Marcetis in rebus vestris, neccogitatis!* "

## LETTER XXI.

*Shemniz, Sept.* 13, 1770

YOU have seen by the series of letters, which hitherto I have written to you, that the mountains in the *Bannat, Hungary* and *Transsylvania* consist of granites, clay, lime, horn-and sandstone. But which are the most ancient? which are the richest? which in each of them is the rule of the veins and fissures? how have the different rocks succeeded each other? These queries properly answered might furnish a compleat theory of the *Hungarian* mountains.

The most ancient mountains in *Hungary* and its dependencies can only be observed in the highest mountains; and even there care is to be taken against the illusion of the superincumbent outside. The *Carpathian* mountains, for example, might be considered at first sight as calcareous mountains, if examined only in such places where mines are working in slate and limestone; but that would prove as wrong as if a man was to fancy that hills covered with vegetable mould are thoroughly composed of it. Examining with this caution the *Carpathian* hills as far as the *Marmaros*, and those which separate the *Moldaw* from *Transsylvania,*

*vania*, and rife near *Werfbez* in the Bannat, we conftantly find that their undermoft ftrata, or rather their main-bulk and nucleus, confifts of granite.

I had ordered a young man, ufed to gather foffils in the *Carpathian* mountains, to obferve every rock appearing above the fuperincumbent mountains, and to bring me famples from thofe rocks as well as from the higheft *Carpathian* fummits. I had them, and it was grey granites. You remember the granite-chain running towards *Konigfberg*, which I mentioned in my laft. I obferved there, that the granites is in the hading, and argillaceous rock in the hanging fide. Hence it appears that the argillaceous rock is accumulated and fuperincumbent on the granites.

The granite rocks in the mountains behind *Alt-Sol*, running to *Upper-Hungary*, the fame rocks which I mentioned at *Felfo-banya* and *Kapnik*, and run there underneath feveral fuperincumbent ftrata to the *Carpathian* mountains, prove to conviction that they confift of granite. Look over my letters again from the Bannat. You will find noticed feveral places, in which the granite appears peeping from under the flate and limeftone. Mr. *Delius*, in his Treatife on the Mountains, has likewife mentioned fome fuch rock; and thefe are granites, which includes the *Hazeg* valley, and feparates the *Moldow* from *Tranffylvania*. It appears there either entirely naked, or covered with a flate

or

or limestone roof. All these facts together are fair evidence of the higher *Hungarian* hills consisting of granites. It is material to add, that there is no place in *Hungary* in which the granite appears to be naturally incumbent on other rocks; that wherever it appears above ground, the superincumbent more modern stone strata are easily to be distinguished; and that granites has never been found in any mine to alternate stratified with other rocks. I know by the mineral-history of other countries that the same has been observed in general. However, I would not be understood as did I suppose the inner part of the globe to consist of granite. It is rather probable that granite in such depths, which hitherto we have not reached, and perhaps never shall, are accumulated on rocks of a simpler mixture. Nevertheless granite is the most ancient rock hitherto observed; and that opinion is greatly confirmed in *Hungary*.

To my knowledge there have not been hitherto found any metallic veins in the *Hungarian* granite-mountains. The *Althandler* vein at *Konigsberg* runs in the separation between the granites and the metallic-rock. Perhaps after-times will lay open these hidden treasures. That veins are running in other countries through the granite is a fact too obvious to want here to be evidenced by me.

The

The second species of rock, which seems to have been produced after the granites is argillaceous, such as hornslate or argillaceous mica, thoroughly mixed with quartz. 2. *Kneifs*, consisting of quartz, mica and lithomarga. 3. The metallic rock, being a hardened clay, mixed either with quartz or sherl and spar or lithomarga. And lastly, 4. trap and shivery clay.

Hornslate is to my knowledge very scarce in *Hungary*; but you have seen by my letters, that the lead-veins near *Modern* in the county of *Presburg*, on the foot of the *Carpathian* mountains, run in hornslate: and at *Ruskowa* in another lead-vein, under the chamber of *Shemniz*, likewise on the foot of the *Carpathian* mountains, the common rock is of the same kind: But these veins are leaping, thin and inconsiderable.

*Kneifs* is on the sole of *Simon Judas*, at *Dognazka* in the Bannat. Between *Saska* and *Moldova* whole mountains consist of it, but unobserved to contain any copper-vein; and near *Shemniz* the *Kaiserstolln* in *Hodriz* is driven in *Kneifs* for an auriferous quartz vein.

The most common and the most remarkable of these argillaceous rocks in *Hungary* is the metallic rock, which I have described before. Near *Konigsberg* we have found it immediately incumbent on granites. The large and constant gold and silver veins at *Shemniz* and *Kremniz*, as likewise the many rich veins at *Felso-Banya, Kapnik, Nagy-Banya, Nagyag, Fuses, Boitza* run in it. In the Bannat the constant copper-veins, nay even the richest mine at *Dognazka* are found in the same rock.

Our *trap*, which I have seen only in a single mountain at *Kapnik*, contains but small veins.

The *argillaceous slate* is the common rock at *Schmolniz, Newsohl, Tshavoja* behind *Kremniz*, in the Bannat, and in the salt-works at *Torda, Marmaros* and *Sovar*. It contains commonly short, thin and leaping copper-veins, running either across or along this rock under incumbent limestone.

*Limestone* is the third, and if we do not take notice of the accidental beds, the most modern species of rock. In the Bannat it is constantly accumulated or deposited on clay. In the *Oravitza* mountains some pretty constant copper-veins, and at *Saska* and *Moldova* some short copper-veins are working in it. In *Transsylvania* behind *Nagyag*, from *Darsha* to *Glut*, I found calcareous hills incumbent on argillaceous rocks, in which some poor copper-veins baffeting out to the day were working. I do not remember to have heard of any metallic veins in the many high calcareous hills, which are superincumbent on the granite-hills

hills in the *Carpath* or those which separate the *Moldaw* from *Transylvania*. From all these facts it follows, that veins between slate and limestone have constantly a hading of slate and a hanging side of limestone. I am really of this opinion, though I have had many objections against it, when in the Bannat I found some mines with a hanging of slate and a hading of limestone; but on nearer examination these my scruples disappeared for very good reasons, which I will speak of in the description of the accidental rocks. Here I am obliged to take notice, that lime is very often immediately incumbent on granite. Thus I have been told, at least by many people who have seen the *Carpathian* hills, and may be very well accounted for by supposing granite rocks uncovered by argillaceous rocks when the limestone beds were produced.

Thus far of the three ancient species of rocks, known in *Hungary*. Though it be impossible to determine whether they have been produced within a short space of time, or whether they have been accumulated in a long succession of many ages, I am rather inclined to the first opinion. I have seen granite, whose surface, where it was in immediate contact with the incumbent clay, was entirely changed into clay; nay I keep in my cabinet some pieces of granite, with inclosed

fragments

fragments of flate. I have feen argillaceous mountains, and noticed them in my letters from the Bannat, which by the fuperincumbent lime were penetrated and changed into marle. Might not thefe facts incline us to believe, that the granite and clay-beds were ftill in a ftate of a wet pafte, when the fuperior beds were accumulated and depofited upon them? and that accordingly the origin of thefe different rocks cannot be greatly diftant in refpect to time? But I go aftray in hypothefes which in this place are to no purpofe. It is a great fatisfaction to me, that my obfervations on the origin of the rocks agree with thofe of the beft naturalifts. Being eftablifhed on experiments and facts, which I have feen myfelf, I am under no apprehenfion, that my fyftem can poffibly be confidered as an adopted opinion, or what is ftill worfe, as a fancy hatched in my ftudy. Baron *Haller* has given in the preface to his Defcription of the *Helvetia*, a fine account of the *Alps* in *Switzerland*. He is very explicit that the higheft tops of the *Alps* confift of a rock, which is compofed of glimmer, quartz and a loofer ftuff, probably feldfpath, is of a granite-fpecies, and goes in *Switzerland* under the name of *Giefbergerftein*. The common *Alpine* rocks are a fpecies of flate; and the lower hills are covered with limeftone, fome forts of marble,

marble, and other hard rocks. The same is confirmed by Mr. *Gruner*, in his account of the ice-mountains of *Switzerland*. Lord *Bute* has noticed the same rule in the *Pyrenean* mountains, and communicated that observation to Baron *Haller*. The *Tyrolian* mountains are granite covered with slate, as I see by a collection of fossils, which Baron *Adolph Meyer* has brought me from that country. The same is observed in the *Bohemian* mountains. When I lived at my country seat *Altzedliz*, on the high mountains which separate *Bohemia* from the *Upper Palatinate*, I examined these hills with attention. The whole chain of mountains, which from *Bavaria* runs to the circle of *Eger*, is granite, in several places covered by hornslate and other argillaceous rocks. Near *Eger* and *Mautdorf*, towards the *Palatinate*, on the sloping of these mountains the first limestone-hills occur to the observer. Baron *Pabst von Ohain* at *Freyberg*, and Messrs *Charpentier* and *Lommer*, professors in the miner-academy at *Freyberg*, have made several excursions and observations to the same purpose in the *Harz* and *Saxonian* mountains; and the *Swedish* are of the same nature according to your observations and those of Baron *Linneus*. It is to be hoped that naturalists, skilled in mineralogy, will henceforth examine this opinion wherever they should happen to have an opportunity to do it; in order to bring it to fyftematical evidence, as highly interefting to philofophers and miners.

But let us return to the *Hungarian* mountains, and examine their accidental rocks. Such are in my opinion fome limeftone hills, the fandftones and fome flate-ftrata.

It is a hard tafk to determine, which limeftone is more ancient, and which accidental. The greateft mineralogifts afcribe the origin of lime-ftone to a deftruction of marine fhell-fifhes. But is it poffible that thofe immenfe maffes of lime-ftone fhould be owing to the animal kingdom? A great part of the *Carpathian* mountains, the bordering hills of the *Moldaw* and *Tranffylvania*, the mountains in *Steyermarck*, and a great many more to my knowledge are almoft entirely covered and buried under limeftone. What immenfe quantities of fhells would not be requifite to the origin of thefe limeftone-hills? *Cronftedt* has obferved already, that the granulated and fcaly limeftone is deftitute of petrifactions. Are we by that intitled to rank thefe fpecies under the ancient calcareous rocks, and thofe with petrifactions to the accidental ones, produced by more modern inundations? I fhall leave that entirely to the determination of thofe learned men, who have more principles and obfervations than I have acquired

quired myself. But it is fact, that in *Hungary* hitherto no metallic veins have been discovered in rocks filled with petrifactions. I rank however the stalactite-like limestone beyond *Altgeburg* and *Newsohl* under the accidental limestone. In the same class I rank the sandstone, which in *Hungary* surrounds the nobler metallic mountains, as near *Nagyag* and *Facebay* in *Transsylvania*. Sometimes they appear in insulated hills; and often they are accumulated on calcareous ground. In these sandrocks there has not been hitherto found any constant or metallic vein. The *accidental* slate is often accumulated on this sandstone, on lime and other ground. It is produced as the stalactite-like limestone by rain or other water, which washed and carried together the dissolved particles of the more ancient mountains. Thus it covers, for example, the coal-beds between *Kremniz* and *Shemniz*, near *Roniz* and near *Waizen*. By the same reason it appears as the hanging side in some mines of the Bannat of *Temeswar*. I saw at *Saska* a mine whose hading was limestone, and whose hanging was slate. The copper-vein was in the limestone. The old hanging side, if there was any before, seems to have been carried away by accident; and then the dissolved parts of the higher argillaceous hills, which I mentioned in my journey from *Saska* to *Moldova*, have been

P 2 carried

carried there and composed the present hanging. The red slate at *Nagyag*, near *Boicza*, and near *Zalathna* in *Transsylvania*, seems to have had a similar origin.

As indurated clay is not so easily dissolved by water as limestone, this slate never occurs in large beds or in considerable hills. The mixture of dissolved limestone and argillaceous parts has probably produced the marle, which in respect to its œconomical uses is entirely neglected in *Hungary*.

It is certainly matter of surprize to you, that hitherto I have not mentioned the hornstone: Petrosilex: But I freely confess I am at a loss where to rank it, whether among the old or the accidental rocks. In my letter from *Zalathna* I described the mountains at *Facebay*, especially the *Loretto*-mine. This rich gold-mine is working in hornstone, incumbent on argillaceous beds; but it clearly appears to be produced by modern inundations, as you will remember from the letter to which I referr you. The petrifactions in the white hornstone near *Lehotka* prove that this species of rock belongs to the accidental ones. I cannot consider the horn and flint-stones as produced by the gelatinous substance of marine insects; a fancy which once a good naturalist has in confidence entrusted me with, but I rank it

next

next to the current hypothesis, of limestone being the remaining substance of destroyed and dissolved sea-shells.

The veins in these hornstone-mountains are more constant and richer than those between argillaceous slate and limestone. Those at least, which I described to you in my letters, contain gold and silver-ore. If this rock should belong to the more ancient ones, which I leave to your determination, it must have been produced at the same time as the calcareous strata were produced, since I find it never incumbent on lime but constantly on clay. Perhaps future observations will prove some species of hornstone, like some species of clay, slate and lime belong either to the ancient or to the accidental rocks.

All these primogenial and accidental mountains and rocks owe their origin to water; and have been produced, either when the world was raised from the chaos, or according to Mr. *Linneus's* opinion, when the whole earth was covered with water, and the precipitation, crystallisation and dissolution of so many animal and vegetable substances brought forth so many new stratifications; or finally they arose from later inundations.

I should here mention those mountains that are produced by fire. There are actually some marks of such mountains in *Hungary*. The vitreous black lava, *pumex vitreus Linnæi*, at *Tokay* in *Hungary*, and several sorts of lavas from the *Carpathian* mountains, give conjectures and evidences of that kind. But to be particular on that account, I want to examine the whole chain of the *Carpathian* hills, which I have a mind to do next year if my impaired health puts no stop to these my intentions.

LETTER

## LETTER XXII.

*Shemniz, Sept.* 28, 1770.

BY the last post I received an order from the court, to accept of the vacant commission of Count *Colloredo* in the board of mines at *Prague*. I do not know whether I shall rejoice at it or not. It is out of my power to visit the *Carpathian* mountains; however I am to follow where destiny calls me to no unprofitable situation, and I am preparing for my journey to *Prague*. Though I have but a few moments to spare, I give, according to your desire, an account of the different ores found in the *Lower Hungarian* mines.

Native gold is very scarce in the royal mines at *Shemniz*, though in general the ores are auriferous, being for that very reason pulverised and washed. In the beginning however of January last, they have found in *Emperor Francisci Stoln*, in the field of *Siegelsberg*, and in a soaring fissure, which runs towards the great *Bieberstollner* vein, a lump of sound red silver ore, mixed with glass ore, and covered

covered with some native gold. A hundred weight of this ore contained about 1270 ounces of silver. In the private mines the native gold is more obvious; it appears in a capillary form on quartz, in soft and brittle glass ore on the *Hofer* and *Antony de Padua* mine. So it is found likewise on red silver ore. At *Kremniz* and *Konigsberg* it is still more common. At *Kremniz* it is often found in lamellœ. I have likewise a fragment of irony quartz with native gold from *Ladislai-Stolln* at *Bugganz*. The *Lower Hungarian* gold is in general to be cleared from auriferous wash-ores or from zinnopel, which is a mixed red jasper, containing gold, silver, lead, zinc, and pyrites. It is the common rock of the *spital* vein, and in general strikes fire with steel, though there are some looser species of this zinnopel, which taint the fingers and look like red bole. Perhaps its constituent parts will prove it to be of the bolus kind; the looser species appears often in a globular scaly form like button ore striking on harder zinnopel: If found stratified in wash-ores, consisting of blende, lead, and a blueish clay, it goes under the name of *string zinnopel*, (*schnur zinnopel*). Mr. *Scopoli* is at present about a laborious chemical analysis of the
zinnopel,

zinnopel, which he intends to publish in his *Anni historico-naturales.*

I can form no idea of the yellow zinnopel, mentioned by Mr. *Cronstedt*, unless he means to give that name to irony jasper. If that should be the case, we might as well give the same denomination to the red jasper, which is so common in the *Calvariberg* and the *Pacherstolner* vein where it is rejected as rubbish. The denomination of zinnopel implies an auriferous quality.

Another species of auriferous stamp or wash-ores is the irony quartz found at *Bugganz*, and in the *Kaiserstoln* at *Hodriz*. You remember perhaps a passage in the account of mining works published at *Freiberg*, in which is conjectured, that irony quartz is generally auriferous; our stamp ores seem to confirm it.

The pyrites, separated from the lead and blende by pounding and washing, contains likewise a great deal of gold. A hundred weight of this pyrites yields fifty-four pounds of stone or lech, and three denarii of silver, which per mark contains fifty denarii of gold. The pyrites at *Konigsberg* and *Kremniz* are still richer. Pyrites containing silver, go here under the name of *Gelft*.

Native silver is still more uncommon in the *Lower Hungarian* mines; all the while I have been here I have got but two samples, one from *Old Anthony*

*Anthony de Padua Stoln* and the other from *Therefia-Shaft*.

On the first the native silver appears in long and thin threads, like human hair, sticking on quartz. The second is a pale yellowish pyrites, from which the silver seems to be grown; this is the more precious for me as *Henkel*, if I am not mistaken, denies native silver to be ever found on pyrites. We are the better furnished with other scarce silver ore.

*Glass ore crystallized* extremely scarce. The miners call it here *Weich-Gewaechs* or soft ore, in order to distinguish it from the brittle glass ore. You have seen in my cabinet glass ore in cubical forms from *Siegelsberg*, and another knotty species from *Moderstoln*.

*Glass ore*, brittle, called here *Roesh-Gewaechs*, is silver mineralized with more sulphur. It contains from four to five hundred ounces of silver per hundred; often its value is but seventy or eighty ounces. Its description in *Justi's chemical works* is exaggerated and extravagant. *Scopoli* will probably give a better account of its constituent parts.

*Red silver ore* is found at *Shemniz* and *Kremniz* either found or crystalized. That found at *Kremniz* is auriferous. On *Old Anthony de Padua Stoln* near *Shemniz*, I have met with dendritical red

red silver ore on white quartz; and on *Rudaina Anna Stoln*, between *Konigsberg* and *Shemniz*, I found it light coloured and sticking in pyrites. Dr. *Moller*, at *Newsohl*, has in his cabinet dark red silver ore in globular forms. *Scopoli* is likewise analyzing this sort of ore.

*White silver ore* is very common at *Kremniz*. It is auriferous, and commonly as it were an incrustation of white quartz. The miners at *Kremniz* call this incrustation *Blachman*; but those at *Shemniz* give this name to the pyritical incrustations of glass ores or rocks, and it is constantly observed near the richer ores.

*Grey plumose silver ore*, from *Old Anthony de Padua Stoln*, different from the *Saxonian* species by its being cast in white quartz, not in capillary crystals but in star-like spots. There is a large vein of this ore. It takes a good polish, which pretty well sets off its star-like form, and the silver sprinkled in antimony.

*White plumose silver ore.* I am of opinion, that this species is no where else to be found. Some years ago it was very plenty in *Old Allerheiligen* mine at *Hodriz*. Its white crystals resemble the white crystalline pin-like horn ore, and stick in a matrix of irony quarts.

*Goose dung ore*, of the same form as described by *Wallerius*, *Spec.* 301, No. I. of a
yellowish,

yellowish, green, and reddish colour, was dug in considerable quantity at *Windish-Leuten* near *Shemniz*; one hundred weight yielding only 800 ounces of silver.

*Silver ocher*, of a brownish, yellow, and white colour found in the same place; containing from three, six, to fifty, and one hundred ounces of silver; the native silver often visible in it.

*Field-spath*, containing silver; of a yellow, red or white colour and hard contexture. Roasted in fire its colour changes to brown and black; but then the sprinkled quartz particles appear to sight unchanged in their colour. It contains from four to eight ounces of silver; is found concomitant with richer-ores on *Siegelsberg*, *Christinashaft*, and other private mines. *Brunnich*, in his supplements to *Cronstedt's Mineralogy*, §. 35. has noticed already the blue colour, which appears on some species of spar in *New Anthony de Padua Stolln*, and constantly indicate a richer silver value.

*Blende*, containing silver. I do not know the globular ore, which *Cronstedt* mentions, §. 175. However, such a species of blend may have been found in former times, which were remarkably negligent of such curiosities. Nevertheless it is fact, that our blende constantly contains some silver though in a scanty quantity; and for this very reason, it is never thrown away among the

the rubbish, but stampt like other ores. It is commonly brown, solid and of a scaly contexture; however, I have found here knotty, black, yellow, green, semipellucid, whitish and several crystalline species of blende.

Lead ore, contains silver, and is commonly of a granulated or lamellous contexture. However, there are likewise several sorts of crystalline ore. White and grey lead-spar is found at *Windisch-Leuten* in the above silver-ocher. Blue lead spar found in the same place.

Copper-ore is found with other metals in the *Spitaler-vein*, but the greatest plenty in *Herrngrund* near *Newsohl*. It consists of yellow and grey copper-pyrites, fallow-ore, and copper-green.

Iron-ore is digging near *Roniz*, *Thaisolz*, and *Libeten*. Commonly it is yellow and blue hæmatites (*Cronstedt*, §. 203.) Black button-ore is considered as something rare. Such is the dripped ore from *Boinik*, in the surface covered with points or pins two inches long, each of them incrustated with blue chalcedony.

Quicksilver never occurs in a native state; but cinnabar-ores appear now and then, though in no such plenty as to deserve parting. If found concomitant to richer ores, they contain some denarii of gold. During my stay in this place they

have

have been found on *Siegelsberg*, on the *Windshaft* and on *John's Kluft*, commonly in white loose clay.

Antimony was found last year in *Three-Kings Stolln* on white quartz, formed like stars. It is however scarce in the *Shemniz-mines*. But at *Kremniz*, in the *Rothish-mine*, noble samples of crystallized antimony are dug out. Sound antimony with native gold, though scarce, found at *Magurka*. Red antimony said to have been found in former times on *Althandle* at *Konigsberg*. The scarce antimony samples, I am possessed of, are found at *Konigsberg*. One consists of fistulous antimony covered with a red incrustation; the other consists of accumulated antimony-crystals, each covered by an incrustation of white quartz. *Cronstedt* mentions a similar species.

Arsenic never offers in the *Lower-Hungarian* mines in its semi-metallic or calcareous form: For this reason our miners are less subject to diseases than those in *Bohemia*, *Saxony*, and the *Upper-Harz*. There has been however found between *Kremniz* and *Newsohl*, in a bed of grey sandstone, red arsenic of a fibrous texture.

Sulphur is found mineralized as pyrites in many different forms, as capillary, globular, undulated, dripped and crystalline. I have mentioned the

the orpiment of *Thajova* in one of my former letters.

Vitriol drips into ftalactite-like forms in the *Pacherftoln*. There is plenty of it every where in the *Old-man*, the roof and the drifts. Its colour white, green, yellow and brown. At *Herrngrund* near *Newfohl* it appears often in blue and rofy-coloured ftalactites; the laft fpecies now and then mixed light blue. It generally contains within fome moveable water-drops.

*Halotrichum Scopoli*, or the hair-falt, feems to me to be vitriol. I do not fee that it is materially different from vitriol. It bloffoms on the fides of the galleries at *Shemniz*, *Kremniz* and *Newfohl*.

I have no time to fpare for a defcription of the many different ftones and earths, which I have collected here. I am poffeffed of an innumerable variety of quartz and fpar-cryftallifations; which you fhall find defcribed in the catalogue of my foffils, intended for print as foon as I fhall be fettled at *Prague*.

My next from *Vienna* will tell you what hope of recovery is left me, and what curiofities deferving your attention I have met with.

LETTER

## LETTER XXIII.

*Vienna, Oct.* 19, 1770.

YOU are acquainted with the diversions, advantages and disadvantages of this city, and the state of learning in this capital university has not escaped your observation. You complained in one of your letters, that, among so many expensive establishments for the sciences, a profession and a collection of Natural-History has been most unaccountably forgotten. If among those, who are intrusted with the reformation and improvement of the sciences, a single friend or connoisseur had hinted it, her majesty would not have neglected it, as her royal care and munificence has amply exerted itself in so many ornaments and improvements of the university. Unluckily *Van Swieten* is neither a remarkable friend nor connoisseur of Natural History; a deficiency easily to be pardoned in a man, who is so eminent in many other parts of learning.

With your observations in hand I examined the imperial cabinet of fossils. You have scarce left me any new discovery. However, you have

over-

over-looked a great fragment of black vitreous lava found in *Hungary*. It is thrown in a corner, as deserving no great attention. They are the more particular and forward with their pretended gold-grains inclosed in raisins, with their gold incircled vines, and the gold threads, supposed to have grown as plants. But all these rarities are downright impostures. Yellow resinous sap is looked upon as gold-grains, and the pretended vegetable gold threads appear to an unprejudiced eye, what really they are, artificial gold-wires. I will allow that they have been found twisted around the vines; but might not these remains of ancient *Hungarian* dress and magnificence have been hid in the ground, by accident have been torn up with the vines, and by error have been considered as vegetable productions? This is the more forcible, because these vegetable gold rarities are generally found near *Tokay* and *Altschl*, places renowned in history for having been residencies of *Hungarian* princes and kings, and equally known for many battles fought in that neighbourhood. Even the *Hungarians* of the present age delight in saddles, harnesses, swords, and weapons, ornamented with massy gold thread.

The collection of fine and precious stones is really admirable. I was remarkably pleased by
the

the diamonds, half white and half red, or half yellow and white.

Though it is impoffible to form an exact idea of the whole by a fight of a couple of hours, I did not however find here either the compleat varieties of minerals, nor the infenfible gradations of varieties of different ftones, nor any of thofe mineralogical fingularities which diftinguifh even at firft fight the cabinets of a connoiffeur from thofe of mere collectors. So I miffed likewife the greater part of the fcarce minerals of the imperial ftates.

Mr. *Jaquin* has gathered in *Hungary* a fine cabinet of foffils. Have you feen his native gold in molybdæna from *Rhimazombat* between *Newfohl* and *Schmolniz?* The botanical garden under his infpection is likely to be very foon the firft in *Europe.*

The collection of the *Minories* refembles rather to a raree fhow than to a cabinet of natural curiofities. I go very often to Baron *Moll.* His chofen collection of minerals, which is fo remarkably rich in fine petrifactions, and his explications, give me both inftruction and entertainment. Pity it were if this fine collection fhould be feparated fome fome day or other; but this feems to be the fate, fince his fons have no inclination for this part of fcience.

The

The collections in the academies for noblemen have afforded me great delight, not indeed by their rarities, but by their infpiring young noblemen with fome relifh of this fcience. Every noblemen in the *Therefian* college has a fmall cabinet of minerals, fhells and infects in his apartment; and *P. Shiffermuller* fpreads his tafte for Natural Hiftory among thefe young people with great fuccefs, and very good hopes for aftertimes. This learned man is to publifh the butterflies of *Auftria*. The *Piarifts* in the academy of *Savoye* and *Lœwenburg* have of late eftablifhed Profeffors of Mineralogy, and they think ferioufly of encreafing their collections.

At *Prague*, I fhall be almoft entirely deftitute of literay company. Mr. *Peithner* is the only man, by whofe fcience I may improve my knowledge. It is unhappy that we are doomed to live in fo diftant countries. If I were fo free as you, my impaired health fhould put no ftop to a trip to *Carlfcrona*.

THE END.

# ABSTRACT

OF

Mr. J. J. FERBER's

MINERALOGICAL HISTORY

OF

BOHEMIA.

Publifhed in GERMAN at BERLIN, 1774

# ABSTRACT

OF

## MR. J. J. FERBER's

MINERALOGICAL HISTORY OF BOHEMIA.

Publifhed in GERMAN at BERLIN, 1774.

Catharinaberg, *in the Circle of* Saaz.

THE mountains about *Catharinaberg* confift of gneifs, which is a mixture of quartz, mica, and white or reddifh half petrified clay. This clay has been by fome mineralogifts confidered as lithomarga; but it is commonly deftitute of its qualities as defcribed in *Cronftedt's Mineralogy*, §. 78. It is rather common clay. *Terra porcellana phlogifto aliifque heterogeneis minima portione mixta*. Cronftedt, §. 78. 2.

Between *Catharinaberg* and *Grunthal* detached columnar bafaltes, common on the highway, tumbled from the adjacent hills, in which it feems to be incumbent on gneifs.*

---

* Between *Lowofiz* and *Topliz* the mountains confift generally of granites, in which red feldfpath is predominant. It is ftriped and undulated with blackifh glimmer. Columnar bafaltes ftands on fome of thefe granite-hills. I have feen thereabout bafalt-rock, deftitute of regular prifms, but confifting in a large mafs, cracked and fplit in many pieces, more or lefs angulated, and containing plenty of black fherl cryftals.

The gneiss-mountains at *Catharinaberg* are continuations of the metallic mountains at *Freiberg* in *Saxony*, which consist of the same rock. They run towards the *Saxonian* upper metallic-mountains, and insensibly degenerate into argillaceous slate, as may be seen even at *Marienburg*. On the *Bohemian* side the same degeneration of gneiss is to be observed in the further run of the mountains; but it continues here longer in an unaltered state as far as *Joachimsthal*, where the veins are found in argillaceous slate, and even that extremely micaceous wherever it bassets out.

For these reasons, and in respect of its situation and extent, gneiss is to be considered as a variety of argillaceous slate; and in respect of its mixture it might, with as much probability, be considered as a variety of granites, instead of feldspath mixed with clay.

No body will disapprove of these assertions, as chemical and other observations have proved that mica *(glimmer)* is produced by clay, and resolves again into the same substance; that part of the substantial earth of clay is flinty; that clay changes into quartz and other flinty stones, which by art can be reduced again to clay in an aluminous form; and finally, that quartz and feldspath, by the action of air and age, dissolve into a white clay,

clay, for which reason many granites contain this white clay in the place of quartz and feldspath. If this clay should not be considered as dissolved quartz or feldspath, but rather as their original earth (as now and then seems to be the case) it makes no material difference. However, the dissolution is here more probable, since in the circle of *Pilsen* in *Bohemia* many hills of granite, of pure argillaceous slate, of grey micaceous gneissslate, and of hornslate, are observed to be affected by the air; so that their outside, for two or three feet, is changed into a white and clayish substance, which, in the granites, scarce offers any visible mark of their former constituent parts of quartz, mica, and feldspath; nay often they loose their very hardness and stony concretion, so as to appear dissolved and mouldered into a white, loose and soft clay, in which but a few mouldering quartz and feldspath grains, with some mica lamellæ, are to be distinguished; the latter changed from their black and glossy brightness into a pale silver-colour. Few of these stony granite-particles are visibly remaining in this clay; there are however enough to prove its being a solution of granites, which is the more easily to be granted as granite in itself is composed of argillaceous

sub-

substances. ‡ Similar argillaceous solutions of the pure, micaceous and gneissy clay-slate are very common in the circle of *Pilsen*. This white clay is dug for example near the new inn in the neighbourhood of *Ostraw*, near *Innichaw* and in several other places; and it is made use of in common

‡ Having in the year 1753 visited and examined the *Blocksberg*, which is the highest mountain in the *Harz forest* and in *Germany*, I found it in its whole extent and wide spreading ramifications, consisting of grey granite. Where this rock rises above the metallic slate of the *Harz-mountains*, it appears, either entirely naked, or more or less covered with swampy combustible flaw turf, produced by rotten vegetables. On the very highest summits of this famous mountain and on its wide branches, such as the *Little Blocksberg*, the *Heinrichshohe*, the *Bruchberg*, the *Rennekenstein*, and many more, the granite appears in immense shattered masses, confusedly piled up—Vast ruins of a former world. In the deeper valleys it appears sound and stratified. The air has visibly affected its hard substance, in changing the colour of its outside, in lessening its hardness, where most exposed to the inclemencies of the weather, nay, in dissolving it into more or less coarse sand and clay. Large beds of this *granite-sand* have been washed down in the valleys, on the slope of these mountains, and to the foot of the most exposed mouldering rocks; nor is there want of *granite-clay*. As the mica in these granite decays is often of a yellow brass or white silver-colour, times immemorial these decays have been considered by the inhabitants of these wildernesses as gold-and silver-ores. The colour of these sands and clays, or pretended ores, differs according to their different solution,

mon pottery, nay, on account of its white colour, it is employed as lime in the wafhing, whitening and incruftating of houfes and walls. If this clay be found in fuch beds, as baffet out or are expofed to the day water, then it is commonly mixed with heterogeneous matters, and its colour is accord-

folution, mixture and pofition, as Mr. *Ferber* has very juftly obferved. I found on the higheft fummit of the *Blockfberg*, near a rock, which is called the *Devil's-Chancel*, a fine pale yellowifh clay, and a reddifh fpecies in another place, which imparted to the hands, when rubbed with it, a fine filver-or gold-glittering. This vifibly derives from the fine folution of the mica, and, together with the remarkable faponaceous foftnefs of the clay, forcibly caufes me to fay fomething of the finer *China-clay*, as having the fame qualities. 1. I know that a very fine and white fpecies of *China-clay* has been difcovered of late in a vein, which croffes the granites in the *Bruchberg*, connected with the *Blockfberg*. I know, 2. that the *Petuntfe* of the *Chinefe* is a more or lefs decayed granites. And 3. that the *Kaolin* of the *Chinefe* has been confidered by many as the fubftantial earth of granites. Therefore I fhould be inclined, and think myfelf intitled to conclude, "that *China-clay* is but a fine folution of decayed granite ; that there is a good chance to find *China-clay* in or near any granite mountains ; nay, that perhaps fuch a clay may be produced by proper artificial decompofitions of the granites."

Whether the quartz or other fine fand, which covers fo large and extenfive parts of the world, and of the fea-ground, may be afcribed to diffolved granite-mountains or not ? is a queftion which I cannot pafs filent. The many detached
granite-

accordingly brownish or yellowish. Sometimes these colours may be owing to a stronger irony-mixture of the rock, or to the yellow brownish clay-flate, which is not uncommon in these parts. These sorts of clay, the white as well as the yellow, turn red in fire, the white less than the yellow, which evidences its being less impregnated with iron than the latter. The yellow clay is used as common loam in walls. In a great part of the circle of *Pilsen*, and on the frontier mountains towards the *Upper Palatinate* and *Bavaria*, the vegetable mould, or the upper strata of the ground, are extremely loamy. Unless they be accumulated by river-inundations, they may be properly considered as produced by solutions and decays of the adjacent granite and other argillaceous hills, which will, I am confident,

granite-pebbles and rocks, which are so frequent in these sandy plains, seem to favour such an opinion; and as among the pebbles of the sea-shores, besides these granite-pieces, a great variety of the harder remains of the more ancient mountains, nay of different secundary ones, are to be found, such as jasper, porphyry, various coloured hornstone, quartz-lumps, gneiss, hornflate, marble, limestones and flint, which prove that nothing stands proof of age and time; this opinion gains some credit, and this the more so, as hitherto no sort of rock has been discovered, in which a greater quantity of visible quartz be contained. The pure parasitical quartz, deposited in the veins, would prove, I think, insufficient to cover half the world with sandy plains. (Transl.)

account

account likewife for the fimilar nature of other countries.

In thofe places where the granites is not entirely diffolved into clay, but mouldered only to pieces, a brownifh fand is produced, fimilar to that which occurs on many fhores, and which is perhaps owing to the fame caufes.

But now the queftion arifes, how nature operates thefe argillaceous folutions of the rocks? No body indeed will queftion the co-operation of the air and the long feries of ages; but how acts the air? If I was allowed to recur to the general acid of the air, I fhould have done very eafily with my explication, fince I have feen the fulphurous acid in *Italy*, iffuing from the old volcanos, change even the black and vitreous lavas into a white and aluminous clay. But the vitriolic acid in the air begins of late to be controverted, unlefs vitriol-works and other vitriolic exhalations account for its prefence. However, I cannot help obferving the fimilarity of thefe folutions of lavas, and fuch rocks as contain quartz and feldfpath; and accordingly I am of opinion, that nature operates them in the fame manner. It is fact and experience that air foftens any rock whatever; why or how? that I leave to future examinations. But as the rocks, I am fpeaking of, turn to white clay, it feems to me owing to the acid of their own argillaceous mixture, fince by Mr. *Beaume*'s

excellent

excellent Treatife on Clay it appears, that clay confifts of vitriolic acid, connected with flinty or vitreous earth, and according to Mr. *Poerner*'s annotation, with fome phlogifton. Now fuppofing thefe rocks or ftones to be fucceffively foftened or loofened, and the acid of their fubftantial mixture by humidity or other caufes to be fet in motion; the fame phænomenon ought to arife, which the acid fteams produce in the lavas. Vitriolic pyrites, after having undergone feveral elixiviations, continues to produce vitriol, when expofed for fome time to the air. The reafon is, not, what the ancient chemifts fuppofed, that it attracts new acid from the air, but that its own ftill fixed and unactuated acid is by the air fet at liberty, in motion and in activity, to feparate from the phlogifton, and to corrode the metallic earth. This phænomenon is called the *mouldering (verwitterung)* of the pyrites; and the fimilar alteration of the above rocks goes juftly under the fame name. Their acid forfakes the phlogifton or other mixtures, which fixed it; it fpreads now in the loofened ftone, acts upon the vitreous earth, unites with this earth and produces clay.

But I have deviated too much from my fubject, and return therefore to the gneifs-mountains at *Catharinaberg*. I faid that they are to be
confidered

considered as continuations of the *Freiberg* gneifs-mountains, and of the *Saxonian* and *Bohemian* argillaceous flate-mountains; or, which is the fame, that the whole tract of the *Bohemian* and *Saxonian* mountains, which confifts of gneifs and clay-flate, is the fame ftratum. Befides I have endeavoured to fhew, that the difference of clay, flate and gneifs is not againft this affertion, fince their conftituent parts are fubftantially and really the fame; and that the fame fubftantial earth, which in the upper metallic mountains produced clay-flate, has under other accidental circumftances and mixtures produced gneifs in the lower metallic mountains at *Freiberg* for example, and at *Catharinaberg*.

This gneifs and argillaceous flate is in *Bohemia* and *Saxony*, as in many other countries, incumbent and accumulated on granites, and is in feveral places covered with limeftone; which fully confirms the obfervation, that the prevailing and general rock-ftrata in the greater and higher *European* chains of mountains confift of three different forts of ftones. The loweft and undermoft and moft ancient, which in the higheft tops appears bare and naked above ground, is granites; the fecond fort accumulated or incumbent on the granites is clay-flate, gneifs or fome other argillaceous rock; the third is limeftone. Thefe ge-

neral

neral and more ancient strata of our part of the world are covered with several beds, which are of a more modern date, and consist of clay, calcareous earth, marle, sand, or that slate which lies in flats over the coal-beds, and is never to be confounded with the slate of primitive and older mountains, in whatsoever degree they may resemble each other in their composition and substantial earth. The physical observation of the mountains, of their situation and beds, of the different time of their origin, of their connexion and their various accumulation is carefully to be distinguished from the chemical and mineralogical examination of their constituent parts. A new stratum may very often consist of the same earth or rock as the lower, more general, and more ancient ones; and notwithstanding the similarity of their substance, they may have been produced in very different times. On the contrary, it seems to be consistent with the nature of things, that the same stratum, whatsoever extensive it be, consist of the same rock, if produced at the same time, or if this rock be found changing in a certain distance, that then it consists of such rocks that are similar in their constituent parts, as gneiss and slate. For this reason I have endeavoured to shew their similarity; and notwithstanding granites consists of the

the same constituent argillaceous parts, it does not belong to the same stratum but to a lower and more ancient one. Many remarkable facts in the following description of the *Bohemian* mountains will demonstrate it; but it would be matter of an extensive work if I should attempt particular accounts of all the facts, which evince in *Bohemia*, that granites, clay slate, and limestone are constantly in that situation which I have indicated before. It would engage me to give a general view of the *Saxonian* and *Bohemian* metallic mountains, and of their run in the plains on both sides. That would require a particular physical geography of all the mines at *Johan Georgenstadt*, *Joachimsthal* and *Platte*, and of all the vast mountains between *Prague* and *Dresden*. I have hinted here only a fact, which I could not pass over silent. The evidences will appear in the sequel, and may perhaps be given some day by me, if abler men should not prevent me from doing so. But being again astray I beg leave, before I return to *Catharinaberg*, to obviate some objections, which might be perhaps opposed to this theory.

Though it is an undoubted fact that slate is accumulated on granites, and limestone on slate; though it is likewise fact that these three sorts of rocks are the moſt ancient, moſt extenſive, and moſt conſiderable ſtrata of the world, as far as we have explored its depth: there are however many other more modern, thinner and accidental beds, which either cover ſingle parts of the above primitive mountains, or are accumulated in the valleys and gutters, or on the ſlope of the higher mountains. We need not believe nor pretend that granite ſhould be every where covered with ſlate or limeſtone. No; there are in the contrary many places, in which it appears bare above ground, and in which it riſes to the ſide of the adjacent accumulated and ſuperincumbent ſlate or limeſtone-mountains. In the ſame manner ſlate appears often naked above the limeſtone. It is indifferent whether this have been ſo from the beginning or whether this may have been produced by inundations, earthquakes and other accidents, which have taken away their incumbent roof; nor does it alter at all the above rule of the three ſorts of incumbent ancient rocks.

Let us ſuppoſe that a granite-mountain, within the verge of the greater mountain-chains of *Europe*, be covered by limeſtone, this calcareous ſtratum ought to be conſidered as produced either in later times by a particular and partial inundation, or if really it ſhould belong to the third general and old limeſtone rock, the granites is to be ſuppoſed having appeared naked above the
ſlate

slate before this general limestone bed was accumulated; which very well accounts for the imaginary difficulty of the limestones being found now and then immediately incumbent on granites. The pre-existence of the granites before the slate, and that of slate before the limestone, cannot be controverted, since they are incumbent on each other. No architect can lay the roof before having laid the foundation.

Before I can go on I have still a previous remark to recommend. The examination of the physical origin, and of the common substantial parts of several rocks, shews their great similarity and their insensible degeneration, as I have evidenced above by the example of clay-slate, gneiss, horn slate, granite, quartz, mica and feldspath. Hence in mineralogy, and the description of mines, arises an indispensable necessity to distinguish by constant names, the different degrees and varieties of the rocks. The name of *gneiss* should never be given but to the rock, which visibly contains the three above substantial parts of its mixture; and that variety, which contains only mica in grey petrified clay, and which is destitute of quartz, should constantly go under the name of *grey micaceous slate*. I allow this to be a violent distinction forced upon nature, since the three substantial parts of gneiss are not constantly visible, (which is the case at *Freiberg*) and grey micaceous slate is very often found with gneiss in the same mountain; for being nearly related to each other, I cannot help wishing that the utmost care be taken in their denomination. Mineral bodies, I know very well, are not differenced by nature into different classes and families, as plants and animals; they are but different varieties and different mixtures. Nevertheless it is better to distinguish by words and intellectual distinctions things, which can be distinguished, than to confound them and to be in want of proper expressions.

The mines at *Catharinaberg* are in the *Stadtberg*, which in its length runs between hour nine and ten of the compass, has a valley on both sides, and is of about 340 fathoms diameter. Commonly the veins run in a direction which is parallel with that of the valleys and the mountains; but in this place three noble veins, and several insignificant fissures, run a-cross the mountain in hour two of the compass. These are the *Nicolai*, the *Calves head*, and the *Elizabeth* vein.

The *Nicolai* vein dips somewhat sliding or slipping, *(tonnlegig)* § has no distinct side-skirts *(saalbænder)*

§ The dipping of veins and fissures is determined by their angles and inclination to the horizon, and accordingly measured

*bænder*) and is almoſt inſenſibly blended with and caſt in the mountain ſides. The vein-rock is commonly gneiſs as the mountain; but now and then it conſiſts of a ſpecies of granite. The gneiſſy-mountain-rock breaks and ſplits parallel to the dipping of the vein; but if veins between hour ſix and nine unite or croſs the main vein, the gneiſs next to the vein gets or affects a ſituation, which is parallel to the dipping of the croſs-vein. Wherever the mountain-rock turns harder and ſounder, the vein turns ſmaller and thinner. It is commonly not above one foot wide. A reddiſh irony clay ſoaks through the vein and incruſtates it. This commonly appears in the neighbourhood of richer ores. Fiſſures from the hanging

ſured by a quadrant. The *German* miners give them different names, which are expreſſive of their different inclination to the horizon.

*A vertical or ſtanding* vein dips or falls from ninety to ſeventy-five degrees.

A *tonnlegig* vein dips or falls from ſeventy-five to forty-five degrees; and has this denomination from *tonn* or *tun* and *legen* or *lay*, implying, that *tuns* or *caſks* or *barrels* laid on the hading of ſuch veins, ſink by their own weight to the bottom, and that accordingly ſhafts might ſtill be ſunk in them, which is a great advantage to the works. For this reaſon I ſhould not ſcruple to call theſe veins *ſliding or ſlipping* veins; as that denomination anſwers the idea of the *German* miners.

*Flach-fallend* veins dip from forty-five to fifteen degrees; and might very well be engliſhed by *flat veins*.

*Schwebende Gaenge* dip under fifteen degrees, and I do not ſcruple to call them *ſoaring veins*, as this denomination exactly anſwers the *German* name.. (Tranſl.

or from the hading, uniting with the vein, improve it. White fine clay with quartz is a forerunner of rich ore ; coarse white clay destitute of quartz forbodes no good, fills at last the whole vein and strikes it deaf. Crossing veins between hour six and nine are pretty common ; alterate the direction of the vein; contain some quartz; are deaf in themselves, nay strike deaf the vein unless it be strengthened by fissures from the east. In this case the cross-veins bring rich ores, which in the hading are commonly richer than in the hanging ; but these richer ores are only to be found within the cross.

These observations of the *Nicolai* vein stand in general for all the veins at *Catharinaberg*.

The ore consists of rich silver and copper-pyrites with blue fluor, blende, copper-glass, copper-green and sometimes with some native silver and copper.

## Commotau *in the Circle of* Saaz.

THE mountains from *Catharinaberg* to *Rothenhahn*, *Commotau* and *Sonnenberg* consist of gneiss, which in many places is fine micaceous and greatly quartzous. Some granite rocks appear in these parts above ground. Near the

the alum work at *Commotau* the gneifs changes into that argillaceous flate, which is commonly found with coals. It is much impregnated with vitriolic acid, and properly roafted produces alum. Impreffions of vegetables are not uncommon in it, and it is to be queftioned whether it might not be confidered as a variety of gneifs, or whether it be owing to a later origin, and to an accumulation in fome valley or floping ground of the more ancient gneifs, which is more probable. However, the aluminous flate at *Tolfa* in the *Roman* ftate belong to the fimpler or pretended primogenial mountains.

This flate is in the open fields put in fquare pyramidal piles, fired and roafted. Once fired it continues burning by itfelf; and is left fo for fome weeks till it is quenched by water. Its black colour is then found changed into red. During a year it is left expofed to the air, and then three times fucceffively elixiviated; each elixiviation lafting about twelve hours. The elixiviated flate is during fix months left in the air, till it be brought to the fecond, and after a fimilar diffolution in the air to the third elixiviation. At the firft boiling of the brine it is mixed with urine, and in the lead pans evaporated to a farinaceous powder; which afterwards is diffolved in frefh water and by two fucceffive boilings brought to cryftallifation.

In former times they boiled likewise vitriol; but it was found unprofitable. The annual produce of alum is about 200,000 pounds, which, on account of the great number of alum-works in *Germany*, sells at a lowered price of twelve florins per hundred weight. The other *Bohemian* alum-works at *Eger* and *Tans* are of no importance.

These alum-works induce me to mention the *Bohemian* sulphur and vitriol-works. Those at *Alstattel* in the circle of *Saaz*, at *Nessaberg* and *Grosslickowiz* are the most remarkable. The pyrites, after its sulphur is extracted by firing, is elixiviated for vitriol. At *Kupferberg*, in the circle of *Saaz*, there might be produced towards 100,000 pounds of blue or cyprian vitriol; but there is no opportunity for sale, tho' the price be lowered to fourteen florins per hundred weight.

### Presniz *in the Circle of* Saaz.

THE mountains consist of gneiss, which is white silver-coloured, bluish or dark-coloured. Detached basalt-prisms shew that basaltes is not wanting. They work here for silver and iron. 1. The silver-mine *Maria Kirchbaw*, is in a vein running to the south, betwixt the hours twelve and one, and containing reddish ponderous gyps-

gyps-spar. In the deeper drifts native silver, and in the uppermost ones, nay to the very turf, other rich silver ores are said to have been found of old. Even this gypseous spar is supposed to contain some silver. It serves as a fluor in the copper furnaces at *Catharinaberg*. To the south the vein is found; but to the north, in the sloping of the mountain, it is shattered and deaf. A vein in which the gallery or drain was driven has united with the chief vein under the shaft, but without improving it.

The works are carrying on rather for the fluor, and for hopes, than for any present remarkable produce of ore.

2. *Orpes iron-mine*. It is a common observation, that flats commonly begin on the foot of simple, more ancient or pretended primogenial-mountains, where they sink under the adjacent plains. However, the existence of flats in the midst of ancient mountains, and in their valleys, appears among several other evidences by the nature of this iron-work at *Orpes*, and sundry others more to the north, which produce the same ore and are under similar circumstances. Partial and accidental inundations, as well as the successive washing down of the decays of the neighbouring more ancient and higher mountains, will pretty well

well account for this phænomenon. ‖ At *Orpes* the undermoſt or loweſt ground is gneiſs, on which a large bed of ſcaly limeſtone is accumulated and incumbent. This is covered by a ſoaring vein or a bed of iron-ſtone thirty fathoms thick. The whole is buried under a white argillaceous ſtone ſtriped

‖ One of the moſt remarkable flats of this kind, which is undoubtedly a marine bed, has been more than once examined by the tranſlator. It is in the *Harz* foreſt between *Zellerfeldt*, *Altenaw*, and the *Calenberg*, in the midſt of ancient mountains, and appears there in ſeveral places near the *Feſtenburg*, the *Schulenberg*, and the *Calenberg*, not only in deep valleys as Mr. *Ferber* ſuppoſes, but even on and near the top of the higher ancient ſlate-mountains. Such is its ſituation on the ſides of the *Calenberg* and the *Schwarzeberg* towards *Altenaw*, and remarkably ſo near an inn, which is called the *Auerhahn* in a quarry known there by the name of *Shalke*. The tranſlator therefore is of opinion, that this flat or marine bed, as many others of that kind, is rather owing to more general revolutions and cauſes than Mr. *Ferber* ſeems to admit; though the partial cauſes, alledged by him, may be without diſpute admitted for thoſe ſecundary beds, which are deſtitute of marine bodies. The ſecundary marine bed on the higher *Harz* mountains is a very inſtructive phænomenon. It conſiſts of fine ſandſtone, commonly white, compact, fine grained and pure; and ſo it is found in the *Shalke*, and is cut into the form of grinding and ſcythe-ſtones. But in ſeveral places it is of a greyiſh or browniſh colour and mixed with ſome fine mica flakes. Such it is commonly near the *Feſtenburg*. It contains a great variety of fine impreſſions of ſcarce and moſt part unknown

striped with mica, to the thickness of seven fathoms.

The above large soaring iron vein, if not rather iron stratum, contains the finest iron-coloured ores, which resemble the *Swedish*; such as *ferrum refractorium mineralisatum, tritura atra, textura chalybea.* LYTHOPHYL BORNIAN, p 124.

*Ferrum refractorium, textura granulata, granis minimis. Ibid.*

known marine sea-bodies, as of *Asteriæ columnares rotundæ perforatæ.*

*Asteriæ solutæ solares,*

*Ostreo pectinites anomius vespertilio alatus,* whose impression produces a species of hysterolithus.

*Ostreopectinites anomius planus latior,* whose impression produces a species of hysterolythus, peculiar to this place.

*Entomolithus paradoxus, trilobus transversim rugosus;* or a new and undescribed species of Cacadu or *Dodsley*-fossil, peculiar to this place.

*Conchæ minores læves*; and *Cochleæ* and *Buccinitæ lævęs minores.*

The decays of this remakable sandstone bed appear many miles on the other side of *Goslar*, in the plain country near *Kloster Grachof,* and towards the *Steinfield*, in detached dragged and blunted sandstone pieces, filled with the same petrifactions which are scarce in themselves, but the more so as included and moulded in sandstone, which is commonly destitute of petrifactions. I forbear in this place to draw consequences from this singular fact; which, obvious in themselves, prove to conviction that very violent causes have co-operated to raise and to destroy our highest mountains. (Transl.)

which are found with sherl, garnets, wolfram, mica, hornblende, grey amianth, green sudflag, (*Cronstedt Mineral.* §. 106. *Wallerii. Mineral.* Edit. 2. p. 398. 5 & 6.) and a deaf irony green clay, in which all these minerals are commonly sticking.

The scaly limestone, which is the hading or the floor of this iron-vein or bed, is grey towards the south. In both places it bassets out.

The white argillaceous stone, striped with mica, which is the hanging or the roof of this vein or bed, seems to be produced by the decays of the adjacent mouldering higher gneiss-mountains. This successive accumulation is supported by several argillaceous beds in the neighbourhood, ¶ and by many pieces of fossil wood, which are penetrated with iron, and are so frequent in the upper flats of the argillaceous ground near *Orpes* that even some good iron is extracted from them.

¶ Near *Presnitz* they dig in a gallery a green painters clay *Cronstedt*, §. 79, and near *Kaaden* white *China*-clay, with a white grey, milky, opaque, argillaceous stone, which is smooth and glossy in its fractures and a production of the *China*-clay, as appears by the loose clay contained in its middle. Mr. *Peithner* has described it in his *Mineralogical Tables* under the name of *Porcellanites*- *China*-clay is found in several other places in *Bohemia*, for example, at *Lumpe*, near *Bochmish Giefkubel*, at *Zitolib*, on *Sonnenwirbel* near *Weyperth* at *Laun*, *Marklin*, *Hlubloss*, &c,

Weiperth

### Weiperth *in the Circle of* Saaz.

THE following mines are working here for silver and cobolt. 1. *Clementis-Stolle.* The mountain rock is gneifs; the vein not opened yet. Where it baffets out they have found under the turf a fpecies of very rich brown filver-ore. It feems to have been horn-ore; but want of knowledge and curiofity has deftroyed it by unconcerned fmelting. The fituation of the ground did not permit to drive the gallery in the vein; they have therefore driven it in the gneifs, but fo unwifely, that they have not thought neither of a place where to put up the rubbifh, nor of the neceffity to include the brook which runs before the entrance, and of courfe overflows it as often as it fwells by fudden fpring-or autumn-waters.

2. *S. Anthony Stolle* is driving in the vein, in a gneifs-mountain. On the fole of the gallery the hanging is flate, and the hading gneifs. This flate is a branch of the neighbouring metallic mountains in *Saxony*, and a variety or degeneration of the gneifs, which appears clearly in this place, where they border together. They have found in this gallery fome filver and cobolt.

3. *S. John*

3. *S. John in the Desart.* The vein runs in gneiss.

## Joachimsthal *in the Circle of* Saaz.

ALL the *Bohemian* metallic mountains, from *Catharinaberg* hither, consist generally of gneiss, which is a mixture of visible quartz, mica and a whitish clay. Near *Joachimsthal* this visible difference of the constituent parts disappears, and they are so closely mixed together, that the rock thence produced is to be called grey micaceous and quartzous clay-slate. It is the common rock of all the mines at *Joachimsthal*, and turns less micaceous in the depth, where it is more argillaceous, soft, lamellous and black, and the more resembling to the argillaceous slate of other metallic mountains, especially those at *Clausthal* in the *Harz forest*. However, the grey micaceous clay-slate continues in some mines to a great depth before it changes its nature.

The metallic mountains at *Joachimstahl* are towards the south of a gentle ascent, run in different ridges to the east, to the north, and to the west (the highest being that which runs to the north) and to the frontiers of *Saxony* sink down again in the plains. The valleys between these ridges are extremely deep; accordingly the hills are remarkably

ably high, which has afforded to the miners a good opportunity to work many galleries, which from every side converge to the south, and to the valley in which the city of *Joachimsthal* is situated.

These ridges are to the south; the *Adelsgreen*, and the *Little Mittelsberg*; to the north-east rises the *Turkner*, and in the west the *Pfaffenberg*; to the east is the *Hoheberg*; to the west the *Upper-Turkner*, the *Shottenberg*, the *Kohlberg*, the *Keilberg*, and the *Under-* and *Upper Niclasberg*. All these hills are very steep, and naked. Their exterior mould is sandy and barren. The facility of working by galleries has induced the miners of old to search the numerous veins from every part. Hence that astonishing number of old galleries. Above forty of them are still working, and many of them, for example the *George* and the *Theresia* stolln are driven a length of 450 fathoms.

All these galleries and works of *Joachimsthal* are divided into six different fields, and belong to the following companies their tenants.

1. *Unity*, belongs to the community of the citizens at *Joachimsthal*.

2. *Hohe Tanne* is belonging to the imperial court.

3. *Hubert* or *Helena-Hubert*.

4. *Friedenfield*.

5. *Schweitzer Gewæltigung*, divided between the court and private companies.

6. *Sæchsish*,

6. *Sæchſiſh Edelleuth-Stolln*, and *Apple-tree-Stolln* at *Abertham*, belongs to private aſſociations.

All of them are drained by two deep canals or levels, *Barbara* and *Daniel*. The former has its door in the city, and including its ſeveral wings is driven 4,500 fathoms. In a direct line it is 1600 fathoms, and its ſole is 170 fathoms perpendicular under the higheſt top of the mountain. It was the deepeſt gallery of the ancients. *Daniel* is, including its wings, driven 5600 fathoms, and in a direct line 1500 fathoms. It runs twenty fathoms underneath the ſole of *Barbara*, empties under the town more to the ſouth. Theſe draining-levels are kept in repair by the court, for an allowance of a ninth part of the profits.

There are but three drawing ſhafts for all theſe mines. *Hohe-Tann* dips ſomewhat ſliding nearly in 73 degrees. Its whole depth is $60\frac{1}{2}$ fathoms. *Unity-ſhaft* is 154 fathoms perpendicular. *Hubert-ſhaft* is 70 fathoms perpendicular. Hence ariſes the neceſſity that theſe ſhafts ſerve by alternate turns to different mines and aſſociations.

The works are every where ſunk much underneath the ſole of theſe ſhafts and galleries, ſo that theſe mines, after thoſe in *Tyrol*, are the deepeſt in the world. They have from 200 to 350 fathoms perpendicular depth under the turf.

Conſidering

Considering the situation of the high and steep hills, which go under the collective name of the *metallic-mountains* of *Joachimsthal*, and which are separated by deep valleys, one would be inclined to think that the ascent, fall, and direction of their rocks and veins must be corresponding with their exterior form. That is commonly the case in other metallic mountains. But here, contrary to such observations, the rock generally ascends from the south, and sinks either to the west or to the north, as in some respect will appear by the run of the veins, which is absolutely unaffected and undisturbed by the direction of the valleys, except that it seems to influence their quickness.

These veins are very numerous; they fall into an indefinite depth, and continue quick and metallic 350 fathoms. In respect of their general run and direction, they are by the *Bohemian* surveyors and engineers divided into *midnight and morning veins*.

The *midnight* or *northern-veins* run from south to north, between the ninth hour and three, dipping from east to west from 54 to 78 degrees, being in general *sliding veins*.

1. Gold-rose hading fissure runs in
 hour — — — 1 : 6 $\frac{1}{4}$ line.
2. Gold-rose hanging fissure runs in
 — — — — 12 : 5 $\frac{1}{2}$

3. Francisca

3. Francisca — — 9 : line.
4. Anna — — — 8 : 7
5. Fundgrube — — 12 : 6 $\frac{1}{2}$
6. Backer-vein — — 1 : 7 $\frac{1}{2}$
7. Hilbrandt — — 12 : 6 $\frac{1}{2}$
8. Geshieber — — 10 : 4
9. Rose, from Jericho, in hour 2 : 3
10. — — hading fissure 1 : 3
11. Joseph — — — 12 : 1 $\frac{1}{2}$
12. Bergkitler-vein — — 1 : 2 $\frac{1}{2}$
13. Schweitzer — — 1 : 2 $\frac{1}{4}$
14. Ioung Schweitzer — 2 : 4 $\frac{1}{4}$
15. Jerome — — 1 : 7 $\frac{1}{2}$
16. Geisler-vein — — 12 : 4
17. Flat-vein — — 12 : 7
18. Mathesi-vein — — 2 : 2

The *Morning* or *Eastern*-veins run from east to west between hour 3 and 9, dipping from south to north from 60 to 73 degrees, being all of them *sliding veins*.

1. Lawrence runs in hour — 5 : 1 $\frac{1}{2}$ line.
2. Francis de Paula — 5 : 5
3. Joachim — — — 6 : 0 $\frac{3}{4}$
4. Susann — — — 6 : 3
5. Kayserthum — — 6 : 3 $\frac{1}{4}$
6. Corona — — — 5 : 2
7. John in the Desart — 6 : 0 $\frac{1}{2}$
8. Ursula — — — 6 : 6 $\frac{1}{4}$

9. Three

9. Three Kings — — 6 : 5¼ line.
10. Hutten plan — — 6 : 4½
11. Morice — — — 6 : 2½
12. Seegen-Gottes — — 5 : 7¼
13. Geyer — — — 5 : 7½
14. Andreas — — — 7 : 2¼
15. Trinity — — — 5 : 7¼
16. Heer-Paukner — — 7 : 0½
17. Fundgrubner — — 7 : 0¾
18. Spathgang — — 7 : 2½
19. Cow-vein — — — 7 : 0
20. Wasserstolln — — 6 : 6
21. Michael — — — 6 : 4½
22. James Major — — 6 : 6½
23. Rosner — — — 6 : 0¾
24. Elias — — — 7 : 0½
25. Saxen-Kerl — — 6 : 3¼
26. George Stoln — — 6 : 3¾
27. Old Saxen-Kerl — — 7 : 3½
28. Tirre Schonberg — 6 : 1½
29. Himmels-Krohn — — 7 : 0

There are besides them many other midnight and morning-veins, either uniting with the former, or running by themselxes; but not having been yet examined they are still destitute of proper names.

In order to give a general idea of the mines at *Joachimsthal* I have annexed a general map in plate, I. The richest veins are among the northern ones : *Geſhieber, Fundgrube, Roſe from Jericho, Berg-Kittel* and *Jerome* ; among the morning ones, *Morice, Geyer, Andreas, Cow-vein, Elias, Old-Saxon-Kerl,* and *George Stolln.*

 The vein rock in the northern as well as the eaſtern veins is aſhgrey, yellow, white or blueiſh clay, argillaceous ſlate, and various coloured but commonly red hornſtone *(petroſilex)* which is the matrix of the richeſt ores. The *Roſe-ſpar* is a calcareous ſpar, conſiſting of accumulated roundiſh and twiſted lamellæ, found in the vein called *Roſe from Jericho.* The midnight or northern veins contain for the greater part a very fine red hornſtone, ſemi-pellucid and of a pleaſant colour. It has not yet been found in any morning vein. As ſoon as this red hornſtone appears the clay breaks off, but preſently returns at the end of the hornſtone. They are conſtantly alternating, but ſeem to be of the ſame ſubſtantial mixture.

 The *Paukner-vein* has not ſhewn any thing yet but ſlate intermixed with arſenical pyrites.

 The *Fundgrube* is either entirely filled up with flintlike grey hornſtone, or holds it in neſts and nodules.

 The width and thickneſs of theſe veins are various, from one inch to two feet. The ſame vein appears very different in its thickneſs ; they are

are often so much compressed that no vein but a simple joint only is to be distinguished. The hardness or softness of the mountain-rocks have a share in it; nay, the mountain and vein rocks turn softer as soon as the vein itself turns quicker and nobler.

The veins do well in general in the ascent of the mountains; and richer ores are ever to be expected in the crossings of the northern and eastern veins. If both veins be filled with clay there is no chance of ore.

If one be filled with quartz or calcareous spar, and the other with hornstone, the ores turn rich in the crossing; infallibly if two veins, one argillaceous and the other hornstone, by dipping cross one another, rich ores are produced.

In these crossings the veins are now and then disturbed in their run and dipping, and if they do not improve they are shattered.

Ramifications or fissures separating from a vein, and uniting afterwards to another, raise its value and thickness, especially if they should happen to unite with it in an acute angle, and continue to run with it for a considerable length. This happened in the *Geyer*, where they have at prefent a profpect of the richeft ores. However, the contrary happens likewife, and veins have been by the uniting fiffures fo much compreffed and fhattered, that fcarce any track has been left remaining; then experience has taught, to trace the run of the deftroyed vein by that fiffure, which contains a thin covering of clay one fingle line thicknefs.

The fiffures of the *Rofe from Jericho* in the hading are fcarce ever worth working, though commonly larger than the vein itfelf.

Befides thefe metallic veins fome deaf ones, of a confiderable thicknefs, go a-crofs thefe argillaceous and metallic-mountains. They are called here *combs (kæmme)* and deferve particular notice. Some of them confift of red *porphyry*, which is called here fandftone, and fome of a fpecies of trapp, which is called here wacke.

The *porphyry* confifts of a red flefh-coloured hornftone *(petrofilex)* and milky feldfpath grains, in which fometimes vitreous quartz-grains may be diftinguifhed. In fome places this *porphyry* is foft and unpetrified. Then the feldfpath grains are caft in a reddifh loam. So I found it in the *Kuhgang*. Some large veins or combs of this porphyry crofs the metallic mountains at *Joachimfthal*, commonly from fouth

to

to north. They unite with the veins, run parallel with them and crofs them, now and then improving their metallic value. Such a porphyry vein is next to the *Cow-vein*; two of them are clofe to the *Schweitzer*, one near *Elias* on *George-Stolln*, but the largeft, and till yet the only improving one, has been found on the *Rofe from Jericho*. Here it has in the hading united with the vein, and has produced the richeft glafs-ores which ever have been dug in this mine, but broke it off with the porphyry. The whole width of this porphyry-comb has not been exploit *broke* it is fuppofed to be at leaft eight or ten fathoms. In its cracks and fiffures this improving porphyry-comb contained a remarkably fat clay; nay the porphyry foftened by it as other vein-rocks turned rich.

The *Combs*, confifting of a fpecies of *Trapp* or hardened irony clay, are commonly of a grey and greenifh colour. Some are black, and in this fpecies white calcareous fpar-grains and greenifh fherl are found in the *Cow-vein in the Unity*. Their run is very regular. Their width and thicknefs from fome inches upwards to forty fathoms, remakably large where they baffet out. They unite, run parallel with the metallic veins, and crofs them either to their improvement or to

their

their various disadvantage, striking them deaf, altering their run, or compressing them so as to be with great pains to be found again. Under ground these combs are often so hard that they cannot be worked but by blasting; but crossed by galleries, or by other accidents exposed to the air, they wither and moulder into that argillaceous earth which formed them, change their colour to yellow and ochraceous, and turn very saponaceous. This clay for the greater part dissolves in water. In respect to their run and direction they are hereabout as other veins called *Morning* or *Midnight-Combs*. The former cross the metallic midnight-veins and dip from north to south. The latter cross the metallic morning-veins and dip from east to west. Some of them are entirely perpendicular.

There have been found above thirty *Morning-Combs* which basset out. One of them *Schon Erz* is thirty fathoms where it bassets out. Five such *Wacken*-veins have been crossed by the works between the *Cow-vein* and *Elias*. The *Rose from Jericho-vein* is crossed on the level of the *Daniel-Gallery* by three such combs, fourteen, nine and eight inches, two of them twenty fathoms deeper have united together, and with the vein, which

has

has been so much improved by this accident, that *Wismuth* has begun to break in the hading.

The *Midnight-combs* or *Wacken* commonly disturb the run of the metallic morning-veins. The *Cow-vein* is crossed by three large combs; two of them are on the *Theresia-drain* near the *Geyer*, under ground from some inches to six feet; but where they basset out they have a thickness from thirty to forty fathoms. The third comb is grey, and on the *Barbara-Sole*, between thirty and forty fathoms. In this place, which is 150 fathoms perpendicular depth, and about 3000 fathoms distant from the door of the gallery, they found of old, that famous antidiluvian tree, which is the more remarkable as it lay in the midst of these slate rocks and the comb I am speaking of. The exterior appearance; the inner stripes or fibres; the concentric circles; the ramification of this substance into round branches; the soft bark which stuck to them; and something like leaves found in several parts of this grey stone, or even on this substance itself; in short, every visible circumstance convinced the first discoverers of their having found one or more petrified trees in the midst of the mountain; and the pious simplicity of these former times, which considered the most natural phænomena as prodigies or signs of divine warnings, dared not consider these

trees

trees but as having been buried here by the deluge, as appears by the name of this petrified wood, and by the different accounts of *Matthesius* in his *Sarepta*, and in his *Chronicle of Joachimsthal*. Soon after this discovery the water prevailing, and this whole drift of the cow-vein giving way, it has been ever since either impossible or extremely dangerous to examine this place, which is so curious for Naturalists. Many mining officers at *Joachimsthal* have since attempted to doubt whether these pretended antidiluvian wood-like blocks ever have been real wood. They have considered them as a fibrous and black variety of the grey comb-rock, which they supposed generally to rise from this place, and thence to diverge in the several ramifications, which cross the mountains at *Joachimsthal*. This last supposition cannot be admitted, as these combs are so very different in their direction and dipping. But even the first supposition is destitute of foundation, unless a man should allow to himself to conclude from the remarkable scarcity of petrifactions, in the simpler or pretended primogenial mountains, that no such petrifactions are to be found or to be admitted at all. A Naturalist cannot help wishing to see these old drifts and caverns cleared again, in order to take fuller information of this singular phænomenon; but hitherto no information is to be had except what

may

may be gathered from the accounts of *Matthesius* and those fragments which are kept in several cabinets. Their similarity with other petrified wood, especially with beech-wood, is so striking, that even the very first sight of it keeps down any doubt which you might have entertained against it, except you should be inclined to look upon the petrified wood in general as *lusus naturæ*. I remember to have seen in Baron *Pabst von Ohayn's* collection at *Freyberg* some very unequivocal samples. It was impossible to mistake the wood fibres, the yearly annular circles, the ramification of the branches, their roundish form and the soft unpetrified bark, which some people have falsely supposed to be amianthine fossil cork (*suber montanum.*) I cannot say any thing of the leaves, since I never did see them; but for my part, I am perfectly convinced that it is really petrified wood, and that I may rather depend upon the evidence of my eyes than upon the objections of some wise pyrrhonists, who might consider even the petrified shells of calcareous strata as sports of nature.

Subterraneous caverns have not been found generally but in calcareous hills. However, I know by very good authority, that a cavern of a remarkable width has been discovered in these argillaceous metallic slate mountains at *Joachimsthal*, in the midst of solid rocks, and in a depth of

250 fathoms. It was hit in 1772 by the fifth drift, driven by the *Hoh-Tanner* company on the *Andreas Vein*. A short time before they had fine ore in the vein from an half inch to three inches thick; but suddenly, when the miners worked a blasting hole in the hading, a violent stroke from under ground forced the bore from their hands, and a flood of water, spouting not only from the hole, but breaking forth from every rock fissure, overflowed the whole drift, and obliged the miners to fly. Soon after the water ceased to break in from the roof a head, but it continued violently to spout from the bore hole to a distance of three fathoms. The engine could not overcome this subterraneous inundation till a second wheel was set to work, and the drift was drained again, which facilitated the going on with the works, and the breaking into a cavern eleven fathoms length and nine fathoms wide. Its roof appeared foul and shattered; its floor was craggy by large rock masses tumbled from above; and it was still filled with water, which made it impossible then to explore its depth. The sixth drift or level of the works has since been extended and driven towards the same place; however, its depth or bottom is still unexplored.

Having giving the names of the several patentee-companies, and of the several fields which they are

are working, I shall here take notice of the veins, which fall within the extent of their works.

The field of *Hubert* is crossed by *Geyers, Paukners, Trinity's, Fundgrube, Baker, Geshieber,* and *Anna* veins.

*Unity* works on *Andreas, Geshieber, Cow vein, Seegen Gottes, Fundgrube, Backer, Hillebrand* and *Rose from Jericho.*

*Hohe Tann* works on *Geyer, Seegen Gottes, Andreas, Rose from Jericho,* and *Elias.*

*Fridenfeldt* on *Cow vein, Rose from Jericho, Schweitzer, Christopher, Joseph,* and several crossing morning veins.

The field of *Sachsish-Edelleuth-Stolln* (or the *Saxonian* gentlemens gallery) is separated from the former fields, and situated in an argillaceous slate-hill, called the *Dirnberg*, which is independent of, and divided by a valley from the other metallic mountains. It is the highest top to the east of *Joachimsthal*; has its own veins, which seem to have no connexion with those in the metallic mountains. Being emptied towards the day, the works are driven already above 100 fathoms below the level of the valley. Its veins are likewise divided into eastern and northern veins.

The *Northern* or *Midnight Veins* are,

1. *S. Thomas*; runs in hour ten three lines, dips in fifty degrees to the east. It has two ramifications, which

which turn deaf where they unite. The vein-rock is white calcareous spar, red hornstone (*petrosilex*) and blende.

2. *Margaretha Vein* runs in hour eleven six lines; and spreads in two ramifications. They are metallic as far as they are yet pursued. The vein-rock is clay, and argillaceous slate sprinkled with pyrites.

3. *Hulf-Gottes-Vein* runs in hour eleven five lines; dips in seventy-five degrees, and spreads its fissures and ramifications in the hading, which from twenty to twenty fathoms unite again with the main vein. The ramifications separating from it make it deaf, carrying the ore along with them in the hading; but uniting again with it they make it quick and fair again. This seems rather to indicate two veins constantly and alternately separating and uniting. The vein-rock is argillaceous slate, clay, pitch-blende or black jack.

4. *Wolfs-Vein* runs in hour nine and three lines; dips in seventy-five degrees; is never quick but in crosses.

5. *Daniel* runs in hour nine and five lines. Is worked out.

6. *Newheusler* runs in hour ten, is worked out likewise.

7. *Zeitler* runs in hour twelve, dips in eighty degrees.

The

The *Morning Veins* crofs the above northern ones, and are, 1. *Reichs Stollner Vein* running in hour five dipping in feventy-five degrees. 2. *King Saul* running in four and five, dipping in eighty-five degrees; both having in former times produced very rich ores, and are on that account worked out.

I fhall only take notice here of the moft remarkable ores found in thefe feveral veins at *Joachimfthal*, as Baron *Born*, in the printed account of his cabinet, has accurately defcribed all their varieties, and the vein-rocks in which they are contained. But previoufly I am to obferve, that all the ores of thefe veins are deftitute of vifible coverings, nay that often they appear fprinkled in the hading and hanging rocks, though the veins in themfelves be not immediately grown to the rocks, but rather feparated from them by thin argillaceous joints. To fave the ore which is fprinkled in the rock fides of the vein, they are cut down on both fides of the vein one foot thicknefs, and delivered to the wafh-works.

The moft remarkable ores at *Joachimfthal* are as follows:

1. *Native filver* in different vein-rocks; in *Shard Cobolt (Scherben-Cobolt)* and on the black wacke it appears in capillary forms, and is called then *Brufh-ore*. Native filver has been found in *Gefhuber*,

*Geſhuber, Schweitzer,* and *Cow-vein;* though more common on theſe veins it has never been plenty.

2. *Glaſs-ore* is the richeſt ore at *Joachimſthal;* one hundred weight is commonly valued at 180 marks of ſilver. They melt it in lead in order to part it. It is in undetermined, cryſtalline, and grape-like forms, and has been found in former times on *Cow-vein, Roſe from Jericho,* and *Schweitzer,* in ſo large lumps and maſſes, that ſmall pyramids, ſtatues, and many ornamental toys, have been carved of it, as appears from the accounts given by *Mathefius,* and from the many curioſities of that kind kept in the electoral cabinet at *Dreſden.* Large cryſtalline pieces of glaſs-ore, found in former times, are kept for ſale in the archives at *Joachimſthal.*

3. *Red ſilver-ore* found, in undetermined and in cryſtallized forms, is found on *Andreas, Geyer,* and *Backers-Veins,* in arſenical cobolt; in *Trinity* its matrix is arſenical pyrites; in *Geſhuber* it is red hornſtone *(petroſilex)* and on *Roſe from Jericho* it is roſe-formed lamellous calcareous ſpar. This laſt variety, which is at preſent very ſcarce, is in reſpect of its ſingular beauty preferable to any other. The cryſtalline, ruby coloured, and pellucid red ore ſticks on, and often in the midſt of the above lamellous roſy-ſpar, which on that accident reſembles now and then to a roſe or ranunculus, and has given the name of that flower

to

to the vein in which it is found. A connoisseur will, by the very colour of the red ore, guess the mines in which it is found. The *Bohemian* red ore is remarkably fine ruby coloured and pellucid; that from *Andreasberg* in the *Harz-forest*, is somewhat darker, on account of its stronger mixture with sulphur; and iron that from *Saxony* keeps the middle between these varieties. *Agricola*, in the tenth book *de Natura Fossilium*, is of opinion, that the red ore from *Joachimsthal* on the *Barbara vein* is auriferous; which by the assayers is denied. I had no leisure to try it myself.

4. *White silver-ore* is said to have been found in former times on *Andreas* and *Rose from Jericho* in pyrites.

5. *Lead-Glance*, containing some silver, found in former times on *Geyer*, and in 1730 on *Cow-vein*, in one foot thickness. Such poor ores are generally scarce in the veins of finer ones.

6. Yellow copper-ore and pyrites, said to have been found in *Seegen Gottes*.

7. *Cobolt* in different argentiferous varieties occurs with several silver ores in the *Geshuber*, *Hillebrandt*, *Rose from Jericho*, *Schweitzer*, *S. Esprit*, *Emperor Joseph*, and other veins. The pure cobolt, destitute of silver, is stampt and sold afterwards to several cobolt-manufactories in *Bohemia* and the *Empire*. A hundred weight sells from

T thirteen

thirteen to forty-five florins. In former times, the demands of that commodity being then lefs, they produced in the whole Kingdom not above 200,000 pounds; at prefent they produce about 1,000000 of pounds. In the moft ancient times cobolt-ores were by ignorance thrown amongft the rubbifh; for this reafon the bing-places at *Joachimfthal* are fearched over at prefent, and fome wafh-works after the *Hungarian* principles are to be fet up. The greater part of the *Bohemian* cobolt is exported to *Holland*. Though the *Bohemian* ores are as good as the *Saxonian* ones, the preparation of the fmalte is not brought hitherto in *Bohemia* to the fame mechanical perfection as in *Saxony*, where the manufacturers are never at a lofs to work exactly the famples demanded. This feems owing to an imperfect feparation of the ores, and to fome ignorance of their nature and manipulation.

8. *Pitch Blende*, containing fometimes three marks of filver, faid to have been found in 1772 on *Geyer's vein*, occurs in fome other mines.

9. *Arfenical ore*, mifpickel, now and then with fome orpiment found with the cobolt and filver ores. In the old works on the *Hubert-vein* a white arfenical calx drips and coagulates into ftalactites.

10. *Cinnabar-ore*, according to *Mathefius* and *Albinus*, formerly found in *Dorothea-vein* in the *Shottenberg* at *Joachimfthal*.

The

The hardness and solidity of the argillaceous slate or of the mountain rock is a great advantage to the works, since in most places they want no timber.

The engines are built according to the principles of those which are used at *Shemniz*, and which have been described by Mr. *Poda*. The same is to be said of their washing and stamping mills.

The history and the former riches of the mines at *Joachimsthal* have been described by Mr. *Peithner*, in the second volume of the *New Physical Amusements*, published at *Prague* in 1771 in Octavo.

## Aberdam *two hours way distant from* Joachimsthal.

THE mines lie on the limits of the deeper granite and the incumbent argillaceous slate, and afford an easy opportunity to be convinced of the granites being under the slate.

Some veins at this place run in grey micaceous slate, and contain silver and cobalt-ores; some in reddish or alternating grey granite, which here, as commonly in *Bohemia* and *Saxonia*, contains tin ore, though some tin-veins are observed likewise to be in argillaceous slate.

*Eva's Apple-tree* and *Jerome* are the chief silver-mines at *Aberdam*. These veins are at a depth of sixty fathoms, and by uniting with red porphyry veins become remarkably richer. I have seen native silver, in wires and in capillary forms, in yellow, brown and black horn-stone (*petrosilex*) with glass-ore; and hair silver in cobolt, both found in *Eva's Apple-tree*.

The *Morrice-mine* and some others are in granite, and are worked for tin.

*Matthesius* and *Albinus* tell that a transparent cinnabar-ore has been found formerly in *S. Lorence* at *Aberdam*.

## Platte *in the Circle of* Saaz.

HERE are three different sorts of mines, silver, iron and tin-mines.

*Zwittermill* is a high mountain two hours distant from *Platte*.

Its eastern part consists of ash-grey micaceous slate or a species of gneiss, in which *Trinity silver-mine* is working. Its western part consists of solid compact hornslate or a mixture of quartz and mica, thoroughly mixed and penetrated with some iron. This hornslate is extremely hard and sonorous as metal. It belongs to the *Corneus Wallerii*. They dig it in a quarry, cut it into the form

form of pestels, and use it hereabout, and as far as *Johan Georgenstadt in Saxonia*, instead of iron ones, as being less expensive and less obnoxious to the tin-ore. For this reason this hornslate is called hereabout *poch wacke*. *S. John Baptist's* vein crosses this rock, and consisting of blackish, more or less micaceous deaf clay slate, its working has been given up. The hornstone in the hanging and hading of this vein is softer than the common *poch wacke,* grey coloured, mixed with mica, and similar to the grey gneiss on the eastern side of this mountain; whence it appears that this hornstone is but a variety of the eastern and general rock of the *Zwitter-mill*.

In *Trinity* mine on the eastern part of this mountain two veins are working, called *Heavenly Blessing* and *Divine Providence*. To the west they are compressed by the hornstone; but they may very likely in a greater depth unite and fall in together by their dipping, and then prove quick and metallic. It is observed in this mountain, that the silver morning-veins, running in hour 7 or $7\frac{1}{2}$ are quick and metallic if standing or vertical; if they dip less than 45 degrees, that is flat or soaring, they are deaf.

2. *Irrgang* or *Labyrinth* is a large iron vein, which all along its run in hour nine, line four, for a space of about three *German* miles, as far as *Annaberg* in *Saxonia*, is worked by several companies.

panies. Each of these companies gives to its field and part of the vein a particular name. For example, *Maria Hulf mine* near *Platte* calls it *Hilf Gottes-vein*. Here the hading is granite, and the hanging slate; and it has been observed to be richest where running between these rocks. Commonly it runs in granite, which often contains sherl; and it gets a hanging of slate in those places where it runs into the incumbent slate. They told me, that this granite-rock contains some wedges of basaltes. Crossing foul and argillaceous veins force this iron vein into a different direction; but constantly it returns to its rule and main direction. It is about four fathoms wide, and yields the finest red button-ore and other argillaceous red iron-ore, which often appears in a thickness of one fathom. Now and then brownstone or manganese is found with the button-ore, and they use it as a flux.

The *Tin-mines* near *Platte* are all of them in granite.

*S. Conrad* is the chief of all, and works the following veins:

1. *S. Christopher's vein*, running in hour 11, points 6, dipping in 72 degrees.

2. *Fresh-fortune's vein*, running in hour $11\frac{1}{2}$, dipping in 82 degrees.

3. *Conrads-vein*, running between hour 8 and 9, and makes a cross with the two former. It dips in 82 degrees.

4. *Christ*

4. *Christ birth's vein* runs in hour 3.

5. *Matthiew'i vein*. They baffet generally out with tin-stone, which formerly has been washed and thus furnished an opportunity for discovering the veins. The rock, in which they run, is reddish and grey granite, sometimes greenish. Their vein rock is loose granite with parallel layers or stripes of zwitter; which diverging from the vein make it extremely thin. Generally it is from one to four fathoms wide; and so it was about the middle of its depth. The deepest pit is eighty fathoms; the vein still quick. White yellow clay or lithomarga, blackish mica, and fine pointed black wolfram or pyrites are found with the tin-zwitter. The wolfram is here in deaf veins, the constant fore runner of tin. The crossings of the veins, and of the smaller vertical or flat fissures running in hour six improve these veins in general.

### Gottes Gab *in the Circle of* Saaz.

IN this place there is at present no mine working except in the *Kaff*, a mountain consisting of micaceous and quartzous clay-slate. They dig here in several pits, such as *John in the desert*, *Tubal-Cain*, iron ore, good loadstones, and rich but irony tin-zwitter. Accidentally they meet likewife with copper and filver-ore; nay there have been found famples of filver, copper, iron and tin fticking in the fame matrix.

The miners fuppofe that an iron-flat or bed is incumbent here on a tin-flat or tin-ftratum; but that is inconfiftent with the height and nature of the mountain, which is a fimple or true gang-or primitive mountain. Several veins, vertical as well as flat ones, crofs it. The latter carry iron and tin, and being flat and foaring have been by the miners miftaken for flats. The vertical veins ftrike them deaf; and they yield lefs tin as foon as they are united with filver fiffures; this feems to be owing to the fame caufe, which in the *Saxonian* metallic mountains produced filver in arfenical veins (and fuch are the tin-veins) wherever they are croffed by irony fiffures. Inftead of cobolt, mifpikkel, tin and other arfenical ore, they yield then filver-ore. Though fuch particular obfervations are far from eftablifhing general principles, and the alchymiftical conclufion, that iron and arfenic produce filver, might prove too bold and too rafh perhaps; I am however of opinion, that fuch obfervations if made with precifion and veracity, are extremely interefting and ufeful, fince they eftablifh not only rules for particular mountains, veins and mines, but may in time lead to a nearer knowledge of the hitherto too myfterious chemical preparations of nature.

I have

I have said that in the *Kaff-mountain* iron and tin is dug in the same pit and in the same vein. The iron here generally appears in its upper part; the tin, which is here extremely irony, constantly appears at a greater depth; and it would be worth while to examine whether silver-ore might not be found in a still greater depth, which might be easily done by a gallery. My conjecture is supported by a general observation in the neighbouring *Saxonian* rich silver mines. The same vein contains there under the turf iron, in a middle depth tin, and at the greatest depth silver; and this seems to be owing to the same cause, which in any point of these veins, if arsenical, has produced silver whenever, as I told before, an irony vein comes across or unites with them. Having already declared my opinion on the importance of this observation and its consequences, and being far from inclining to alchymistical fancies, I hope not to be asked for a natural solution of this phænomenon. I declare freely to be as ignorant of the causes as I should like to know them; what I know of the matter is, that this phænomenon is fact in the *Saxonian* metallic mountains. It is not I alone who have been convinced of it by several observations; some of the most learned miners can attest it, and have convinced me of it by many examples. Nay in *Saxony* they make use of this and other such observations, and apply them

them with succefs to a scientifical and practical working of their mines; and even the annalifts of the *Saxonian* mining places, but ftill more fo the ancient records, prove that feveral *Saxonian* mines, at prefent producing filver, yielded in former times, and in their upper drifts, good tin-and iron-ore. Mr. *Peithner*, who before his prefent preferment to the character of counfellor, taught at *Prague* the fcience of mines, has found the fame obfervation fupported by feveral correfponding ones in *Bohemia*. I have perufed fome parts of his lectures, as penned down by one of his difciples, and in thefe I find the following paffage: " Some of our *Bohemian* mountains, " fuch as thofe at *Platten, Neudeck, Gottes Gab*, and " above the *golden-top*, near the new road to *Berin-* " *ger*, and fome others, confift in their upper parts " and drifts of iron flats or beds," (which as I have already noticed is a miftake of flat and foaring veins) " but in a greater depth they are fol- " lowed by rich tin, nay even by filver-ore."

I return to the mines in *Kaff*. Pyrites here quickens the tin-ore. The tin-ore, after a previous roafting, is pounded and wafhed on hearths, in which operation they make ufe of the loadftone in order to feparate the iron. Though all the iron particles cannot be drawn off by the loadftone, they are however raifed by it, and the eafier wafhed down.

Formerly

Formerly there have been near *Gottes Gab* silver mines, which are said to have produced a dark red silver-ore, hereabout called native brown-ore.

## Bleyſtadt, *in the Circle of* Saaz.

THE mountains conſiſt of grey micaceous quartzous and argillaceous ſlate, which in a mine called *Heerzug* contains red garnets. The richeſt veins run from the weſt; the northerly veins are deaf and cruſh the former. The rock filling the weſtern or eaſtern veins conſiſts of quartz, which they call *gneiſs*, and appears under a found and compact or a loofe and a farinaceous form.

Their dipping is generally vertical, and their width is from three to eight feet. They are ſteady to a depth of 160 fathoms, till the quartz turns deaf and unmetallic. The ore is coarſe glance, and now and then white lead-ore and brown and reddiſh lead-clay. Formerly there was likewiſe green lead-ore. The glance containing ſcarce any ſilver, they pound and ſell it either as lead to the furnaces at *Joachimſthal*, or to the potters for glazing.

Schlackenwald,

## Schlackenwald, *in the Circle of* Saaz.

THE country is rather gently afcending than mountainous; accordingly they till and plow the foil which covers the deeper rocks. Thefe confift of a mixture of quartz, glimmer, and white clay, which fplits into lamellæ, and forms gneifs, or a continuation and variety of the *Bohemian* argillaceous metallic mountains. They have here three different forts of veins.

1. *Lead-veins* with filver; and particularly a quartz vein in the Emperor *Jofeph's* gallery, running in hour eight.

2. *Tin veins*, fuch as thofe in the crofs mine, which produces wolfram, copper pyrites, tin ore, and tin zwitter in quartz.

3. *Tin ftocks*. In fome mineralogical accounts, the name of ftock-works is too often mifapplied to the uniting of feveral veins in the fame run, or to the larger bellies of fingle veins. In either of thefe cafes there is no reafon, why the name of vein fhould be given up for that of a ftock-work. I underftand by *ftock-work*, that native place of metals, which without any regular or determined run to any certain point or line of the compafs, and without any determined dipping

ping, appears rather as a large conical body of rocks, sunk or inserted in the midst of a mountain. Such a cone or lump consists either of pure ore or of a rock, which is more or less impregnated and penetrated with metallic particles, and is constantly of a different kind than the surrounding or including rock. The whole stock is included within a curve line, which may be either an oval or a circular one. These circular out-lines either unite as commonly in the depth or they diverge. In the first case the stock has the form of an inverted cone; in this it hath the form of a truncated and standing cone. Three such stocks have been hitherto observed at *Schlackenwald*, 1. The *Royal Huber*, which is the largest, and still working. 2. The *Stock-shaft*, smaller and worked out. 3. A *Stock* not yet examined, but reputed to be of the same circumference as the second. Their natural condition being the same, as far as hitherto observed, I shall confine myself to the description of the *Royal Huber*.

The rock which surrounds it, or in which it is standing, is *gneiss*. The stock itself consists of granite, or a mixture of quartz, feldspath and mica-grains and lamellæ, more or less penetrated and sprinkled with tin-ore.

The feldspath is reddish or grey; but often in its place appears a white and greenish clay, which
probably

probably might have turned into feldspath, as a decay or mouldering in the midst of the rock and under ground cannot be properly supposed. This last variety of granite, which is commonly the least rich of tin, is at *Schlackenwald* called *grit (greiss)* and its constituent parts, quartz, mica and clay seem to indicate, that it is but a variety of the gneiss, which surrounds the stock. In a mineralogical respect that assertion is pretty true; and the difference of the rocks being lamellous and of the grits appearing in a compact form, is of no importance. However, this difference is very remarkable and interesting in respect of their origin, antiquity and situation.

This grit-stone consists of grains and lamellæ, ferruminated together as the pebbles in a pudding stone. Whatever be its origin, it is in respect of the time and the manner, in which it happened, of course different from that revolution, in which similar constituent parts joined into an uniform, not granulated more compact rock, which is lamellous, and goes under the name of gneiss. Moreover it breaks in with the granite of the stock, without any remarkable separation or bedding; it never occurs in the including gneiss. Accordingly it is a variety of the granite, and is produced with it in the stock, in the same moment and by the same revolution.

The

The granite of this ftock is comparatively with the common metallic mountains to be confidered as the vein; the furrounding gneifs is in the fame refpect to be compared with country rock. But here the queftion arifes: whether this ftock, like other veins and fiffures, be produced after, and in its præ-exifting mountain-rock? Or whether this granite-ftock may be confidered as a top or fummit of the ancient primitive and deeper granite, which was afterwards furrounded, inclofed and covered by the more modern gneifs? Being convinced by many facts that granite is a more ancient and deeper rock than argillaceous flate, gneifs and other fuch varieties; having moreover feen many inftances of the granites high and bare appearing through and above the more modern and incumbent rocks, I am rather inclined to this laft opinion.

1. Becaufe the common theory of the origin of veins by fiffures or cracks produced in the mountain-rocks, when firft they began to dry, does not account for round conical holes in the mountain-rocks, of fuch extent and regular form as the ftocks under confideration.

2. Becaufe even allowing fuch conical holes there remains the difficulty, how they might have been afterwards filled up with granite, which

commonly

commonly is not found as vein-rock, except in its own fissures.

However, to wave the charge of being too much prejudiced for my hypothesis, I leave my conjectures of the origin and antiquity of this granite-stock and its surrounding gneiss-mountain, to the consideration of the intelligent reader. Nay, in support of those, who might be inclined to consider this granite stock as being of more modern date and produced in the more ancient gneiss, I will candidly relate, that the *Royal Huber Granite-stock* in its upper part is 100, but in the depth only 92 fathoms diameter, having the form of an inverted cone, and being for that the more resembling to the metallic veins, which diminish to the depth, and at last disappear entirely.

But there is besides a third possibility; that of the granite stocks and gneiss-rocks being produced at the same time. Of all the before-mentioned this is indeed the least probable.

Whatever hypothesis my readers should have a mind to, they will observe that no conclusion can be drawn from them against the general observation, that the granite in the largest and highest *European* mountains is more ancient than argillaceous slate and gneiss; and that these are incumbent on granite, as limestone and other modern beds are accumulated on them. Though by my own

own, and so many other observations, I am convinced of that, I cannot deny to nature the power of producing under proper circumstances the same species of granite or rock, which it produced in former times, either by water or by fire, as appears by some lava's which exactly resemble the granitello's. It is obvious that I meant not to speak of these supposed modern granites, but of the ancient ones, in which we can be the less mistaken, as nature in its present quiet course, and without some new and great revolution in our globe, is not likely to produce any remarkable granite rocks, or any far-stretching, high-towering chain of such mountains. The argillaceous slate of the chief, middle and richer, metallic mountains in *Europe* is, in a mineralogical respect, the same as that slate, which is bedded in the more modern mountains; nay which is every day produced by the ooze of every muddy lake ; however, it would be extremely wrong to confound them in a Natural History of the Earth, in the same manner as they are confounded in a Syllabus of Mineralogy. Similar beds of clay or other stratified stones, or different only by accidental circumstances, such as that of colour or hardness, or mixture of iron, lime or phlogiston, are very often found in different depths and separated by other strata. A mere mineralogist or collector of fossils would rank

rank the produce of thefe different beds in the fame clafs, nor think of any difference; but the hiftorian of fubterranean geography cannot fail to obferve their differences in refpect of their antiquity, origin, relation to other beds, metallic contents, and influence on the crofting veins. The fine-grained limeftone of the *Alps*, the fcaly or faline limeftone, the calcareous tophus or travertine, the limeftone incumbent on coals, and finally that fpecies, which we fee every day on the *Dutch* fhore, produced by fea fhells, are lime, nay fome of thefe different fpecies agree in the manner of their origin; though in general they are different in this and in many other refpects. A common fyftematical collector will unconcernedly break the flints from their chalk-matrix, in which they are found in *France*, in *England*, and in *Stevens-Klint* on *Seeland*, and moft fyftematically rank the flints with the other filices, and the chalk with the calcareous earths; but a fcientifical naturalift draws from thefe native places, and other concomitant circumftances, conclufions on the origin of the flints. The former examples illuftrate what I have faid of the granite and its varieties; and the latter demonftrate the neceffity and ufe of a nice obfervation of the different native places of foffils.

I re-

I return to *Schlackenwald*. The mixture of the granite is not uniform throughout the mass of the whole stock. Sometimes large stripes of pure quartz are contained in it, and these are remarkably rich of tin-ore, with blue and green fluors, of wolfram, some copper-pyrites and black-lead (*molybdæna.*) In other places the mica is accumulated in large laminæ; in others prevails the feldspath, or, instead of that, a white fat clay. Though the whole stock be throughout sprinkled with tin-ore, it is however more frequent and more accumulated in the parallel stripes, which cross the finer grey or reddish granite.

Some deaf fissures or veins cross the stock commonly in hour three. The ancients made use of them in their works, especially for firing the ore. One vein, called the *Tscherpermacs-Vein*, crosses our granite stock in hour twelve, and contains tin-ore, pyrites and wolfram in quartz. I am ignorant, whether these veins have their constant run beyond the granite-stock in the gneiss, a circumstance which is observed in the tin-stock at *Geyer* in *Saxony*, and might be supposed perhaps in this, as they are found agreeing in so many other circumstances. I am equally ignorant whether this granite-stock, as that at *Geyer*, have a skirt or inclosure (stockscheider) which in that place consists of feldspath, sometimes mixed with

quartz,

quartz, mica or clay. The tin-stock at *Altenberg* in *Saxony* is of a different nature, since inclosed in granite, and consisting of a variety of granite, in which the quartzous particles are prevailing.

The mines at *Schlackenwald* are said to have been working above 530 years. The *Royal Huber* was formerly held by grant by several companies, who worked this stock in common with the Empress-queen; but hence arose very irregular works, which in this place are of old standing, since the upper fields or levels, having been worked out without any respect to after-times gave way before the year 1580, and caused a large gaping fissure, which is still to be seen. However, they went on again with the same irregularity in the under levels, which are said to be above 100 fathoms under ground. The consequence is, that even at present the works cannot be safely examined, and that any jumping on the turf above the works makes the whole ground tremble and shake. Some time after the ocular examination of these mines by Baron *Mitrofsky*, which happened in the year 1743, her Royal and Imperial Majesty has recalled and redeemed the grants, but to no great advantage, since after the irregular ravages of the old man, which have crippled the whole stock, it is almost impossible to introduce any regular work besides, and the richer and larger
tin-

tin-ore cryftals of former times feem to be gone. Formerly they had here that fcarce fpecies of ore, which is known under the name of white tin-ore.

*Beyer in otiis metallicis*, Part. III. p. 169, has given an account of the preparing, uftulating, wafhing and fmelting of thefe and other tin-ores.

To encourage and to facilitate the *Bohemian* tin-mines in general, her Majefty has ordered her mineral trade commiffion at *Vienna* to buy the *Bohemian* tin at a certain fixed price.

The tin-ores of *Schlackenwald*, *Platte*, *Gottefgab* and of many *Bohemian* wafh-works, were formerly fuppofed to be auriferous, and to contain fome filver. But modern affays have proved the contrary.

## Schonfeld, *near* Schlackenwald.

THE tin mines in this place are the moft ancient in *Bohemia*. Of the former filver mines nothing is remaining but ancient records. The *Simon Judas* and the *Crofs* mine have been the richeft tin-mines, and were worked on veins which ran in *gneifs*. The white tin-ore, which

*Cronstedt*, §. 2. 209, has called *ferrum calciforme terra quadam incognita intime mixtum*, was found in thefe mines. It offers either compact in indeterminate forms, in which cafe it refembles to ponderous white or greenifh and fat quartz, or in white or yellow fplendent cryftallifations.


### Græflitz, *in the Circle of* Saaz.

HAS yellow and greenifh copper-pyrites, with green and brown copper ochres in argillaceous and quartzous flate. In the fmelting of thefe pyrites nothing is remarkable, except that they make ufe of white fluor, fetched from *Saxony*, and that to fome foft pyrites they mix limeftone.


### Mies, *in the Circle of* Pilfen.

THE mountains confift of grey micaceous and argillaceous flate, fometimes mixed with quartz. The veins are generally quartz, which now and then is cryftallifed. In a fpecies of thefe cryftallifations I obferved the fame parallel incifions

cifions, which are a character of the cryftallifations from *Hodriz* near *Shemniz* in *Lower Hungary*. The ore is lead glance, which per hundred weight yields from two to four ounces of filver, and towards forty pounds of lead. Yellowifh green hexagonal prifmatic lead fpar has been found in former times. The works are in great decline, and have ever been very ill contrived, which makes their examination not only difficult but dangerous too.

*The coal work near* Wilkifhen *in the Circle of* Pilfen.

IT is but of late that in this place a gallery has been driven into a coal-bed, which baffets out; and they have but juft begun to fink a fhaft upon it. If I were only to write for thofe who in the art and fcience of the mines fee and conceive no further entertainment than as far as they have a profpect of clear gain before them; and if I were not convinced too, that exact mineralogical obfervations, even of poor mines, may prove very ufeful for richer ones, I might have very juftly fuppreffed what I fhall have to fay of this inconfiderable coal-work and its fituation. But as thofe gentlemen will not poffibly loofe their time with my performances,

performances, I shall rather freely submit my observations to the judgment of the few learned connoisseurs.

The mountains, which to the west divide *Bohemia* from the *Upper-Palatinate* and *Bavaria*, consist of granite and of several varieties of argillaceous slate. The granite of the highest mountains hereabout contains large and thick black sherl-crystals. In my descriptions of the mines at *Catharinaberg* I have given already some account of the rocks of these mountains, which connect with the Circle of *Pilsen*, and to the south run into the Circle of *Prachin*, and to the north into that of *Saaz*. In order to determine the situation of the coal-bed at *Wilkishen*, I am to fix upon a point of this ridge of mountains, and that shall be at *Kladraw*, one hour's way distant from *Wilkishen*.

The mountains in this place consist of pure sometimes micaceous and more or less quartzous dark grey or blueish argillaceous slate. *

* If a great quantity of quartz be closely blended and connected with argillaceous slate, it proves very hard and longitudinally fibrous in the fractures; that is to say, it changes into true *horn slate*. *Corneus Fissilis Wallerii Mineral*, Ed. 2. p. 358. 2. I have been convinced at *Kladraw* of this degeneration. Pure argillaceous slate is commonly crossed by quartz veins. Horn slate has no such veins, but instead of them quartz has internally and equally connected itself with the clay and its produce the mica. It is found here in

the

On the road between *Kladraw* and *Haide*, and towards *Pilsen* fragments of this slate are found, some of them regular cubes, others in the form of oblique or rhomboidal pyramidal columns, from some inches to one foot in height; their form as the cryſtal, n. 8. tab. I. in *Linne's Syſtema Naturæ*, tom. III. In the granite mountains, near *Kladraw*, I have found similar rhomboidal and pyramidal columns of grey and reddiſh granite. Their form is conſtant and regular; in

the ſame horizontal or ſoaring ſituation as the common pure argillaceous ſlate; accordingly it is not conſtantly in a vertical poſition, as at *Edelfors* or in other metallic mountains; nor does it appear in the form of waves. It may be doubted therefore with great propriety, whether that vertical poſition hath been its natural poſition from its very beginning. Moreover the poſition, which may be accidental, does not determine the ſpecies; the leſs ſo as even the pureſt argillaceous ſlate, nay many other modern ſtratified ſtone beds appear not only in horizontal and ſoaring, but likewiſe in many other, and even in vertical poſitions. All theſe circumſtances, together with the chemical aſſays, prove it to be argillaceous *(Wallerii Mineralogy.* ed. 2, p. 355, 358, 359, and 364.) and convince me, that it ſhould not be conſidered as a genus, but that it might be very juſtly ranked with the common argillaceous ſlate *(Schiſtus Wallerii.)* However, as it is a variety, I do not controvert its proper name and claſſification in the Mineralogical Syſtems; but in reſpect to phyſical geography it ſhould ever be ſeparated from the argillaceous ſlate.

ſome

some of them the feldspath, in some others the quartz is decayed into farinaceous clay.

But a third species of columns of the same form is to be found about *Kladraw*; and this consists of a variety of the dark grey argillaceous slate, which seems to be an intermediate substance between argillaceous slate and granite. Its substance is darkgrey, blackish and fine micaceous slate; but it is mixed with parallelopiped spots of white and lamellous feldspath, which exposed to the air turn milky and opaque. By this accident it resembles to porphyry or to a variety of granite. Near *Plan* red garnets are found in it, which proves its being a variety of argillaceous slate, because the *Bohemian* garnets are commonly in micaceous quartzous clay slate. Perhaps the rock, I am speaking of, is breaking in the limits where slate and granite border together. It is in my opinion a remarkable phænomenon, that this variety of argillaceous slate with spots of feldspath, nay even the pure argillaceous slate and granite are found in these parts in constant determined and regular columnar forms; and as they are found so in their natural situation in the mountains, it appears clearly that these rocks in their former state of fluidity had a natural tendency to a regular form whatever be the cause of it:

The

The regularity of their faces and angles would induce me to confider their forms as cryſtalliſations; but as commonly the name of cryſtalliſations is given only to regular aggregations of diſſolved ſaline and metallic bodies, I leave to my readers the liberty to look out for a better name. It is fact, that theſe rocks ſplit and break with as much regularity as poſſibly may be produced by any cryſtalliſation. How it happens, that ſo few granite and ſlate-rocks, and thoſe but in certain places, are endowed with that quality, I am at a loſs to account for. ‖

In the neighbourhood of *Kladraw* is another variety of argillaceous ſlate. It is cloſely mixed with quartz and mica; is extremely ſhivery and a true table-ſlate (ardeſia) for which reaſon they break and uſe it inſtead of tiles. The convent at *Kladraw* is covered with it; and I ſuppoſe it is a fair concluſion that this table-ſlate (ardeſia) is not conſtantly found in modern or flat-mountains (flozberge.)

‖ Let us fairly confeſs that we are equally ignorant of the natural cauſes of the ſaline and metallic cryſtalliſations; and that beſides the above-mentioned rocks and ſtones there are many more, which break into regular forms, though theſe have not been hitherto properly attended to by our Mineralogiſts. I gave a hint of that kind in the firſt *Latin* edition of my Syſtem of the Earth, p. 11, 12, 13.

The

The coal-work at *Wilkifhen* is but one hour's way diftant from *Kladraw*, and the road is thence conftantly defcending. In the adjacent fields, and fcarce a gunfhot diftant, many grey granite blocks appear above the vegetable mould. I could not diftinguifh whether thefe blocks were detached pieces, or parts of the granite-rocks under ground; but it is fact, that the coal-bed, with its fkirts of black and rotten flate, is but from three to fix fathom under the turf. Its roof or cover confifts of a mixture of white-grey clay, fome quartz-grains and fome flakes of argentine white mica; that is to fay, it confifts of a loofe incoherent granite, which on that account rather deferves the name of fand, and in fact is called fo by the workmen. The queftion: whence it was produced? is eafily anfwered by the above defcription of the higher mountains near *Kladraw*; and that roof may juftly be confidered as grit and decays of the fchiftous and granite mountains, wafhed down and fucceffively accumulated upon the coal-bed. But what is the coal-bed in itfelf? I fay, it is the continuation of the argillaceous flate at *Kladraw*, penetrated in this by petroleum; accordingly it is not, as commonly, a modern flat, accumulated on the primitive mountain rock at *Kladraw*, but it is
a part

a part and a continuation of that very primitive mountain. §

Some of my readers will, I am confident, find this aſſertion at variance with the common opinion, according to which coal-beds are generally conſidered as conſtantly belonging to the modern accumulated flats (flozberge) and as never making part of the more ſimple ancient or primitive mountains; but as coals are argillaceous ſlate penetrated with petroleum, ¶ I ſee no reaſon why

§ Though the continual ſloping deſcent of the primitive ground from *Kladraw* to *Wilkiſhen* be favourable to this hypotheſis, it wants nevertheleſs better evidence, as on that ſloping primitive ground modern beds might have been accumulated by the ſame revolutions, which have accumulated ſo many others.

However it would be very unfair to deny the poſſibility and ingenuity of the hypotheſis, ſtubbornly to aſſert with ſome Mineralogiſts againſt facts and reaſon, 1. that coalbeds are conſtantly to be conſidered as the remains of ancient foreſts, and 2. that they are conſtantly covered with and accumulated on modern ſtrata. That aſſertion may be ſupported by partial and local obſervations; but they can never make good a general aſſertion as will appear in the ſequel.

¶ Many coal-beds confiſt viſibly of remains of trees, plants and foreſts, more or leſs bituminous, and more or leſs affected and changed by the ſituation in which ancient revolutions have left and brought them.

Theſe are out of queſtion here. Mr. *Ferber* might however have taken ſome notice of them, by telling, that beſides theſe coal-beds, there are ſome others, which are undoubtedly confiſting of bituminous argillaceous modern ſlate. It would not have hurt his hypotheſis.

nature

nature should not saturate with the same substance any argillaceous slate, whether it be of an ancient or a more modern origin; and I am the more of this opinion, as we are far from being convinced that petroleum is only to be found in modern beds. The distinction between the modern flats and the pretended primitive mountains is less material than is commonly supposed. It merely relates to their different antiquity; for the argillaceous slate of the pretended primitive mountains is as well accumulated on granite as every other modern bed is accumulated either on granite or other rocks; and whether this accumulation arrived in the beginning of the world, or long after, is an object of mere conjecture. The different width of the strata is of no greater importance. The pretended primitive mountains are moreover called *Gang* or *Vein*-mountains, because metallic veins are generally found in them; and as these have been formerly working at *Kladraw* there is no doubt, that the mountains hereabout are justly to be considered as primitive or gang-mountains; However, this circumstance alone would not make good the assertion, since metallic veins are likewise found in modern stratified flats; for example the cobolt-work in *Saxe-Saalfield*; the lead-works in *Derbyshire*, and in many other places; for hence it naturally follows, that the mountains

tains produced by water, or by aqueous folution, in refpect of the different periods of their origin and accumulation, and on no other account, can be with propriety divided into old and modern ones, owing their origin to different degrees of relative antiquity. The *ancient-mountains* might, in comparifon to the modern incumbent ones, and their greater variety of thin ftrata, be called with greater propriety *fimple-mountains*. Their denomination of primitive, original, or gang-mountains is by no means characteriftic.

I am to obviate here an objection, which probably will be made to me by thofe who are acquainted with the country between *Kladraw* and *Wilkifhen*, and between *Pilfen* and *Prague*. It is in general flat, and immediately behind *Kladraw* the mountains flope into a plain, which confifts of clay and loam, and afterwards of coal, lime, and other depofited aqueous modern ftrata. Accordingly this whole tract of land feems to be covered by modern beds, beginning at *Wilkfhen*. But I cannot allow that affertion in refpect to the place I am treating of, fince the coal-works at *Wilkifhen* are not in the plain, but in the floping of the higher ancient mountains. Befides, the coals are immediately under the turf, being covered only by loofe grit and decays of the higher mountains, and deftitute of any regular roof of

lime

lime or marle. I had no opportunity to examine the coal-works at *Kotteſhaw*, which are but one hour's ride diſtant; but I am aſſured for the reſt that the whole tract of land hence to *Pilſen* and *Prague* is generally flat and ſtratified, containing ſome coal-beds near *Pilſen*, *Shabrach*, *Berawn*, and limeſtone beds at *Stiez*, which are covered by greyiſh yellow clay. Probably the brooks at *Pilſen* and *Berawn* were larger in former times, and may be ſuppoſed to have depoſited theſe different beds. The ancient mountains behind *Kladraw* however, which conſiſt of argillaceous ſlate and granite, continue uninterruptedly running under ground; for they appear above ground in ſeveral places between *Kladraw*, *Pilſen* and *Prague*. The grey granite-rocks appear on the road from *Kladraw* to *Pilſen*; and thence to *Prague* the argillaceous ſlate, which is either pure, or micaceous, or a horn-ſchiſtus. All this together coincides to ſhew that the coals at *Wilkiſhen* may belong to the more ancient mountains, which are found running under ground far beyond this place. Even ſo far as *Prague* are obſerved ſome ſchiſtous mountains, which riſe high above the ground. The other mountains thereabout are calcareous, and contain petrifactions; ſo that the common coal-beds at *Shabrach*, and the different ſtrata of marle and clay, which are found between the prominent

nent rocks of argillaceous flate, may be juftly afcribed to one or to different inundations. The grey-yellow vegetable mould between *Kladraw* and *Prague* is for the greater part owing to inundations; however it is partly produced by the decays of the argillaceous and micaceous flate-gneifs and hornftone-rocks, which appear above the ground; and this is evident from the mica flakes, which are found in it.

The conclufion, which I am to draw from this, is, that coals are not conftantly found in the modern flats, but that ancient (pretended primitive) argillaceous flate is likewife now and then faturated and penetrated by petroleum. This may be probably obferved in many other places. Should it not perhaps be the cafe of the coal-works in the high and fhaggy mountains of the *Habichwald* near *Caſſel* in *Heſſe*? † There are feveral others, which

---

† For this I refer the reader to my late account of the *German* volcanos, by which it will appear, that this large and fhaggy mountain is produced by many fucceffive ancient volcanic eruptions, and cannot by any means be ranked among thofe ancient granite- or fhiftous-metallic-mountains, which Mr. *Ferber* is fpeaking of. However, as I have no objection againft his hypothefis, and am rather inclined to agree with him, I fhall liften to his call on the naturalifts, and by laying down in a few words, what I know by my own obferva-

which I fufpect to be of the fame nature. However, I wifh, that intelligent naturalifts might examine this my adventurous opinion, compare it with nature, and confirm or refute it by exact obfervations. It is of fome importance to the naturalift, nay even to the miner, fince it accounts and will account not only for the horizontal and foaring, but likewife for the vertical fituation of coals.

Zinnwaldtion of the coal-mine on the *Habichwald*, not only make a fupplement to the above defcription, but likewife to Mr. *Ferber's* affertion.

On the higheft fummit of the *Habichwald*, which confifts of alternating and various beds and rocks of volcanic afhes, lavas, cinders and tufo, accumulated on more ancient lime and fandftone, there is (befides a fpacious plain behind the octogon,) towards *Hof*, a gently floping ground, which falls or runs down into the *Drufelthal*. In the midft of this ground rifes a hill, which is called the Zigenberg, and is incumbent on the lower coal-bed, which has been worked in many directions, and as it fhould feem, furrounds it on every fide.

The ftrata in this coal-mine are, 1. Immediately under the turf, black vegetable mould, mixed with decays of the furrounding volcanic hills, that is to fay, with fragments of hard lava and tufo; three feet.

2. White fine quick fand, as found on the other fide of the *Drufelthal*, and the hill behind the *Pauls-harmitage*, under many tufo and lava beds, in an ancient pit and gallery, called the *filver-well*, and behind the *Sneckenberg* in a fand-pit immediately furrounded with volcanic materials, but not capt with them. One fathom three feet.

3. White

## Zinnwald *in the Circle of* Leitmeriz.

THE mines in this place are but flowly working, if working at all. The country or the mountain-rock is granite of a different mixture. Some large *foaring-veins*, which very improperly

3. White fine clay, as found likewife under the above fand pit; and under a ftratum of bafaltes behind the great water fpout in the garden at *Weiffenftine*; three fathoms.

Digging in this clay bed to the depth of three fathoms, it begins firft to turn grey by ftripes and afterwards entirely black, till it degenerates into coals; which is vifibly owing to its greater or leffer faturation with petroleum.

4. Coals, having in this mine no roof, as being immediately connected with their fubftantial earth, the above clay; three fathom three feet.

The clay penetrated with petroleum, or the uppermoft imperfect coal, is in fome places loofe, and a fpecies of blackifh or brownifh ruble or rotten-ftone, which being an excellent brown or blackifh painters clay, was formerly fold under the name of *Painters clay* from *Caffel*, till the painter *Huchfeld* died who dealt in it, and had kept its native place a fecret. I have found it again, not only in the coal-mine, but likewife immediately under the above fand pit behind the *Snekenberg*, whence the painter *Huchfeld* is faid to have dug it.

Towards the fole the coals turn harder and richer, being then more bituminous and more like the *Scotch* coal; which again proves the former prefence of fluid petroleum.

improperly they call flats run in it, and these are variously cut off, compressed and altered in their direction by other vertical ones; in which case they are pursued and found out again by the small metallic or argillaceous joints, which to the *German* miners are known under the name of *Schleppungen,*

5. The sole of the coals is white sand rock, extremely hard, and striking fire with steel. As hitherto no pit has beeen sunk through this rock, I cannot tell on what ground it is incumbent, whether on volcanic masses, which in respect of the high situation of the coal-mine on the summit of the *Habichwald* is probable, or on limestone, which seems to be the case in respect of the similar sandstone rocks, found on the other side of the *Drufelthal* near the castle of *Weissenstein,* which is in a lower situation than any volcanic stratum of the *Habichwald.*

From these facts I draw the conclusions, 1. that coals are not constantly found in any regular order of successive strata, nor constantly under a roof; 2. that being in this place visibly produced from clay, saturated by petroleum, they may be found in any place or situation where clay or argillaceous slate is to be met with, in ancient simple or modern stratified mountains, as well as on and in volcanic mountains; and 3. that henceforth coals will not be considered as constantly produced from trees, plants and forests, buried by inundations, though many coal-mines have had such an origin.

I cannot conclude this account without taking notice of some spungy coals, which occur now and then in the coal-bed

*Schleppungen*, and have been described by Baron *Swab* in the *Swedish* Transactions, or in his account of the gold-veins at *Edelfors*. The rock of the soaring veins consists of various mixed granite, quartz partly crystallised, tin ore, blue, green and yellow fluor, pyrites and verdegrease. They appear often on both sides inclosed by *saalbands* or skirts of lamellated cat-gold of some inches thickness.

bed on the *Habichwald*. They are found by nodules within the sound coal-bed; are extremely light, and resemble the scorified spungy lava's, found near the octogon on the *Habichwald*. They cannot possibly be considered as cinders or as coals consumed by fire; since there is no reason why they should have been burnt out, and why their heat or fire should not have catched the combustible matrix of sound coals, in which they are contained.

If I were allowed to suppose that the clay-bed on the *Habichwald* was produced, as those in the *Solfatara*, from volcanic ashes, changed by sulphurous acid into clay, there would be less difficulty to account for these spungy honeycombed coals; since then they might be justly considered as spungy volcanic cinders or consumed lavas, together with the volcanic ashes, first changed into clay by sulphurous acid, and afterwards changed into coals by fluid petroleum. Though I am inclined to think so, I leave it however to the nearer examination of those, who are qualified for the task, and who will greatly oblige the Mineralogists by making out, whether any coals are found under the volcanic strata of the *Habichwald*, which, if so, would greatly alter the hypothesis; but it never has appeared so to me. *(Transl.)*

Toplitz,

## Topliz, *in the Circle of* Leitmeriz.

THE mountains from the above place to this, by the way of the *Oak-foreſt*, conſiſt of granite; and the ground is continually ſloping towards *Toplitz*, till at once it ſinks abruptly into the plain.

The country about *Toplitz* conſiſts of gentle *accumulated modern mountains* of clay, lime and coals. It is interrupted by ſome ſteep inſulated higher mountains, which are calcareous, and ſeem to be detached parts of the lower and gentler flats, raiſed by ſome violent accident. This affords an excellent opportunity to diſtinguiſh the difference of the modern incumbent and the more ancient ſimpler mountains, and to obſerve how the former have been accumulated on the latter, when their vallies were large reſervoirs of water, in which their ſeveral ſediments were ſucceſſively depoſited.

The *hot wells* and the *bath* at *Toplitz* are objects too much and too generally known, to want any deſcription of mine. The bath-water has a ſtrong ſmell of *hepar ſulphuris*. They tell here that, at the time of the earthquake at *Liſbon*, theſe

these wells, together with those at *Carlsbad*, decreased, and then burst out with great violence; whence they deduce subterraneous canals, reaching as far as *Portugal*. But the great distance seems not to be favourable to such a supposition; and a similarity of causes may have produced these phænomena in the same moment of time. Next to the town rises a large limestone stratum, and close by are loam pits, worked for kilns. Near the forest-gate is a coal-bed, covered with some fathoms of clay, which partly is fullers clay, and commonly contains lumps of crystallised pyrites. Immediately under this stratum is a stratum of wood-coals, or fossil-wood, some inches thick; and then the coal-bed, which appears four fathoms above ground, and has not been hitherto explored in its whole width. It consists of shivery coals, and many thin beds of clay, from five to eight inches thick. There are no regular works, the coals being dug as from a quarry. The shivery rubbish is burnt, and the ashes produced sold as manure or dung to the farmers.

## Graupen.

THE road from *Topliz* to *Maria-schein* runs over gentle and low hills of clay and lime. The detached granite, which now and then occurs, comes from the mountains on the other side of *Toplitz*. Near *Maria-schein* suddenly rises a large chain of high and steep mountains of gneiss, which include the bason on this side of *Toplitz*. *Graupen*, a tin-place, is at the foot of these mountains, which are called the *Knutler Gezirck*. Ascending these steep mountains I observed an old vein, worked out up to the turf; but found no remains of old bing-places. Two silver-veins, belonging to the community of *Graupen*, are still working; one called *S. Nepomuc*, the other *Silverbeech*, which proves also in this place, that gneiss is generally a matrix of silver-veins, though here it likewise contains tin-veins, which commonly run in granite, and are then not interrupted. By means of a shaft sunk on a vein, they had discovered others, which unite at a small depth, and all of them are metallic. They might, by a draining level of two hundred fathoms, clear them to a considerable depth; and to judge of the bing-places

places and the ores, which I found there, it would be worth their while. I obferved,

1. Sound lead glance with pyrites, and black jack in quartz.
2. The fame with yellow blende.
3. Lead glance, blende and tin-ore in quartz.
4. Sound tin-ore and lead-glance in tin-ftone.
5. Large pieces of cryftallized tin-ore and blende.
6 Fluor with copper-green and azure.

The mixture of leadglance and yellow blende make it probable, that, in a greater depth, the tin will be followed by filver, as is commonly the cafe in the neighbouring mines of *Saxony*.

Thefe veins are called filver-veins, in refpect to the filver which is contained in the lead-ore; and befides them many tin-mines are working in the fame mountain, but to no great advantage. I examined one of them, called the *Sweif*, in which they work on thin *foaring tin-veins* or *fiffures*, which in refpect to their flat and foaring extenfion greatly refemble the larger ones at *Zinnwald*, being like them croffed by many vertical or dipping veins, cut off or alter'd in their fituation, and accordingly worked and purfued on the fame principles.

They are in nothing different from thofe at *Zinnwald*, except in their thinnefs and in their
want

want of skirts, being immediately grown to the sides of the country or rock, in which they are running. They consist in their whole width of solid *tin-stone*, and it is very rare to meet any quartz in them, which makes them bear the expence of working, although their width is scarce ever above half an inch. The beds or lamellæ of the gneifs-rock in the country, on both sides of the veins, are mostly vertical or standing upright. The soaring veins, if cut off or crossed by vertical ones, produce large joints or riders of tin-stone up and down in the vein. It would be worth euquiring whether these vertical veins contain any silver; but the miners in this place do not know nor care for any other than tin-stone. It is remarkable, that the surface of the mountain sinks and falls in the same direction, as the soaring tin-veins are raised or sunk by the crossing vertical veins.

From the top of this mountain the whole country or valley of *Maria-Schein* and *Teplitz* appears before you. It is covered by many gentle incumbent hills or flats, which are interrupted by some insulated conic steep granite-mountains, for example near *Muhlshaw*, where a steep chain of granite-rocks rises above the ground, and shows to the coolest imagination, that the mountains at

*Graupen*

*Graupen* have been formerly the fhore of a fea, which covered the valley of *Maria-Schein*, and above which the infulated granite-cliffs at *Muhlſhaw* muſt have once appeared.

## Muckenburg *or* Muckenthurmel.

FROM the before mentioned old mines in the *Knutler-Gezirck* I purfued that afcending flope of the mountain to another line of old mines, which is called the *Altenberg*. I paffed near *Creuzgang* and *Manfuetus*, which are ſtill working on larger tin-veins than thofe at *Schweif*. Though five workmen are only employed in *Creuzgang*, the dividend however in 1767 confiſted of 1400 florins, which greatly vouches for the riches of the mountain.

Hence it afcends rapidly towards *Muckenthurmel* or *Muckenberg*, where they have in gneifs funk a ſhaft on a large copper-vein, which they confider to be a *ſtockwork*. Here is the higheſt point of thefe mountains, many hundred fathoms above the valley of *Toplitz*; and hence they flope and fink towards *Altenberg* in *Saxonia*. *Gneifs-rock* continued to *Furſtenaw*, where a variety of
granite

granite appeared, which beyond *Geising* at *Altenberg* had its common and natural colour and mixture.

## Ratieborziz *or* Bergstedt, *in the Circle of* Tabor.

THE silver-mines in this place are the property of Prince *Swarzenberg*. They are in gentle hills and grey or blueish clay-slate; in which are observed some fissures of greenish lithomarga, or half-indurated pot-stone earth or bacon-stone. A variety of veins, which cross those mountains, are worked to advantage.

Count *Kienburg*'s works at *Rzemizow*, a vein which is parallel to that at *Ratieborziz*. To the west of this place Cardinal *Migazzi* has a silver-mine. The old mines at *Tabor*, which were formerly so rich of native-and redgilder-ore, have been of late taken up again by Baron *Kesler*. Near *Budweiss*, and near *Rudolph* and *Adamstadt*, are many considerable old mines in different mountains, which are working partly by Her Royal and Imperial Majesty, partly by the citizens at *Budweiss*, and partly by prince *Swarzenberg*.
They

They have of late found dendritical native silver at *Budweifs*.

The mines at *Ratieborziz* or *Bergstedt*, which belong to the prince of *Swarzenberg*, have been taken up again in 1719. There are several of them. 1. The *Chief-mine* or *Haupt-Baw* at *Bergstedt* consists of *Lawrence*, *Charles*, *Michel* and *Nichols*, which are drained by *S. John's* gallery or the deepest water-level. The veins are *S. Nichols* and *Charles*; both northern veins, or running from north to south, and dipping in about sixty degrees. *S Nichols* dips to the east; *S. Charles* to the west. Where richest, they are about three inches; where yielding sprinkled and mixed ore, they are about two feet. The deepest sole is at present seventy fathoms vertical under ground. About fifty fathoms to the west, and in a direction parallel to their's, runs a foul vein twenty-five fathoms wide, which consists of mixed, white blue and yellow clay, and cuts off the crossing ramifications, which diverge from the chief veins in such a manner, that hitherto they have not been hit again beyond it.

2. The under-work near *Bergstedt*, pursues a vein which runs between hour two and three, and has lately yielded rich ore in the *S. Anthony-shaft*, It contains lead-glance with silver, and that sort of specular blende which shall be spoken of

of hereafter. Where the rock softens and is mixed with quartz, there is likewise native silver.

3. *Dorothea*-mine and *S. George's gallery* are to the west of *Bergstedt*, for the present extremely rich, and work on two veins, which have united. One is called *S. George,* runs in hour two; the other is *Dorothea-vein,* which diverges from the former in the hanging runs in hour twelve. This was formerly exremely thin, ran in hard rock, and contained rosy-coloured feldspath, with some lead-glance blende and a little silver; but as soon as a joint with mispikkel came into the hading, the vein produced white-and red silver-ore, native silver and glass-ore. *S. George* has in respect to *Dorothea* an irregular dipping, because it coincides or unites with it in the depth. As long as it continues running by itself, it is so much cut by foul clay-veins, that it produces but little and poor lead-glance and blende.

4. *Old-Woschiz-Hope God's Blessing-Gallery* was quick already fourteen years ago. They work here on a vein, which runs between hour nine and twelve, and dips between forty-five and seventy-five degrees. It is from two inches to one foot wide, and richest where thinnest. It is chiefly quickened by the coming-in of a winding undulating black clay fissure, which appears sometimes in the hanging; then

then it produces fallow, white-and red filver-ore. It is likewife croffed by fome morning or eaftern-veins, which affect it in a different manner; fince, before they crofs it, the vein appears cut off and dead, till, beyond the crofs, it appears quick again in the former run; fometimes the ore and the vein are entirely loft or hid under the crofs-joint, fall out extremely rich underneath it, but difappear entirely above it.

Three hundred and fifty men, the wafhers and fmelters included, have here employment.

The gang or vein, or the metallic matrices in thefe veins are,

1. Fine white quartz partly tranfparent, partly cryftallifed. Some cryftallifations are lamellous, having on the under part fquare regular impreffions, as in the cryftallifations from *Hodriz* near *Shemniz.*

2. White calcareous fpar, varioufly tranfparent and cryftallifed.

3. Yellow tranfparent calcareous fpar, cryftallifed like combs.

4. Reddifh feld fpath, remarkably found.

5. Clay, black, grey, and yellow.

6. Afbeft like cork, in whitifh, thin and flexible lamellæ.

7. Mifpikkel or white arfenical pyrites, quickens the vein.

8. Sulphurous

8. Sulphurous pyrites occurrs now and then.

The silver-ores contained in these matrices are,

9. Native silver, like wire or hairs, commonly on mispikkel or on yellow cubic sulphurous pyrites.

10. Red-silver-ore, found and cryſtallised, transparent like ruby.

11. Glass-ore, found, cubic, and capillary.

12. White silver-ore.

13. Lead glance mineralised with silver, coarse or fine, cubic or polyedrous.

14. Brown or red blende, knotty or cryſtallised, contains much silver.

15. Specular blende, yellow or greenish, lamellous and transparent, occurrs in large lumps, now and then under a cryſtallised form; broken, it consiſts of large lamellæ, which reflect the light like mirrors. It is extremely rich of silver.

The waſh-and pounding-mills have nothing particular. But the above blende, containing much silver, and being for that reason smelted with the other ores, makes the smelting somewhat difficult. The medium of the silver-produce of these mixed ores is from 40, 50, to upwards of 100 ounces per hundred weight.

THE END.

INDEX.

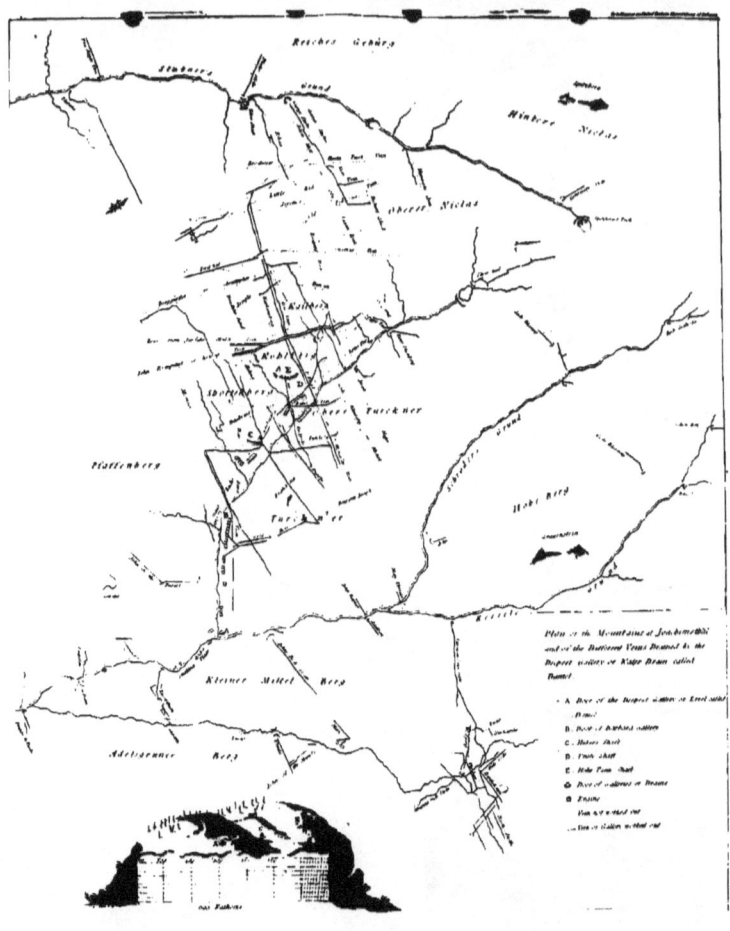

# INDEX.

## A

*ABrud-banya*, a mining-place near *Zalathna* in *Transylvania*, has gold mines, page 117

*Agathe*, white and red, in *Simon Judas* copper vein at *Dognaska*, which runs in metallic rock, 53

*Alabaster* and *Gypsum*, constantly found in and about rock-salt-mines, 144

*Alcaline salt*, native and fossil, at *Debreczin*, 4

*Alum works*, at *Commotau*, 247

*Antimony*, found with mineralized gold at *Nagyag*, 102

————, with native gold, 129, 130, and at *Magurka*, 222

————, grey and yellow plumose; if found in the veins at *Kapnik* improves their auriferous quality, 154

————, grey plumose, in a hanging fissure of an auriferous zinnopel-vein at *Felso-banya*, 160

————, on quartz-crystals, 160

————, radiated, on white pellucid rhomboidal prisms of fluor, 160

*Antimony*, red and grass-green, 160

*Ardellia*, *Wallachian* name of *Transylvania*, 14

*Argilaceous* rocks and stones, *Horn slate*, *Kneiss*, *Metallic rock*, *Schistous clay*, and *Trapp*, in *Hungary* incumbent on granite, 205

and under limestone, which has changed them in some places into marl, 208

*Arsenic crystallised*, or orpiment, found with mineralized gold, 102

———— calx, dripping into the form of stalactites from *Joachimsthal*, 258

*Arsenical*, such as tin-veins, crossed by iron-veins, produce silver in the *Saxonian* mountains, 264, 265

———— veins, such a mispikkel, uniting with silver-veins, improve them, 318

*Bannat* of *Temeswar*, its limits and government, 7—10

inhabitants, 14

mines, 24—27

*Basaltes*

Y

# INDEX.

*Basaltes* columnar prismatic, incumbent on Gneiss and on granite, 228
———, with black crystallized sherl, 228
———, *(grains)* in metallic rock, 33, 34
*Blachmann*, a name given at *Kremniz* to white silver-ore, which incrustates quartz, 219
——— At *Shemniz* it signifies a pyritical incrustation of glass and other rich silver-ores, which it is constantly found concomitant, 219
*Blasting* of the mines, invented at *Freyberg* in 1613; or in 1627, from *Hungary* introduced in *Germany*, 192
*Blende*, with native gold, 129
——— *(specular)* yellow and greenish, transparent; rich of silver; from *Ratieborziz* in *Bohemia*, 320
*Blocksberg*, the highest mountain in *Germany*, and in the Harz-forest, consists of granite rocks, either stratified or confusedly piled up, 231
*Boicza*, rock, veins, mines and ores, 127, 128
*Born (Baron)* in danger to be suffocated in the mines at *Felso-banya*, 158
*Brush-ore*, a species of native silver, found at *Joachimsthal*, 271
*Cassel* (painters clay) described, 307

*Cat-gold*, or yellow laminated mica skirts and includes the tin-veins in-granite at *Zinnwald*, 309
*Caverns* subterraneous, generally found in calcareous mountains; but one of late discovered in slate-rock at *Joachimsthal*, 267, 268
*Chalcedony*, or white horn-stone, with petrified corals, near Lehotka, 194
———, milky, stratified, with detached jasper and agathe, in the same place, 194
———, (blue) incrustation of iron ore 221
———, (blueish) dripped as stalactites on iron-ore, 199
*China-clay*. See *clay*. *Kaolin* and *Petuntse*; a vein in granite-rocks under the *Blocksberg*, 232
———, probably to be found in every tract of granite mountains, 232
———, produced from decaying granites, 232 may be produced perhaps by artificial decompositions of granites, 232
——— and porcellanites or indurated china-clay found in *Bohemia*, 252
*Cinnabar*, solid and scaly, at *Dumbrava*, 120
———, granulated at *Boboja*, 120, 121
———, in a limestone vein, 120, 121
*Clausenburg*,

# INDEX.

*Clausenburg*, ancient *Roman* colony, 146
*Clay*, the substantial earth of mica, glimmer, quartz feldspath and other flints, as mouldering and decaying into clay, 229, 230, —235
———, substantial earth of coals, 305—309
———, produced from mouldering granite, gneiss and micaceous slate, 230—235
———, from volcanic ashes and substances, 309
———, (*auriferous*) with pyrites, in the veins at *Kapnik*, 155
———, (*auriferous*) grey, at *Facebay*, 110—112
———, (*blue*) with auriferous pyrites, 129
———, (*blue*) in slate, a vein producing orpiment, 195
———, (*black*) with auriferous pyrites, 129
———, (*grey*) a vein in slate, with quartz and copper pyrites, at *Smolniz*, 170 —172
———, (*Painters*) from *Cassel*, described, 307
———, (*white*) with auriferous lead-glance, 129
*Coal-bed*, near *Wilkischen*, seems to be a stratum of old pretended primitive slate, impregnated by petroleum and covered with detached decays of granite, 300, 301—305
———, on the *Habichwald*, surrounded with volcanic productions, destitute of any roof, consists of clay saturated with petroleum; perhaps produced from volcanic ashes and cinders, 305
*Coal bed*, at *Toplitz*; common and fossil wood-coal, 311
*Cobalt*, with silver-ore, common in the veins at *Joachimsthal*, 257
———, formerly thrown away in *Bohemia* as rubbish, 258
———, (*grey scaly*) with native gold, 129
*Combs* or *Wacken*, popular names given in *Bohemia* to large vertical veins of porphyry and trapp, 262
*Copper*, cemented at *Smolniz*, 172
———, (ore) *a*. *Azur*, or *blue*, crystallised, &c. in quadrangular oblong truncated forms, 31
B. in polyedrous semipellucid forms from *Saska*, 40
———, *b*. *Brown*; an undescribed species of jasper, mouldering into red copper mulm or tile-ore, from *Saska*, 39
———, *c*. *Broth-ore*; a copper-pyrites penetrated with brown copper-mulm, 31
———, *d*. *Glass*; red, crystallized in triangular and octangular forms, from *Saska*, 38, 39

*Copper*

# INDEX.

*Copper*, (ore) *e. Mulm*; at *Saſka* incumbent on limeſtone, page 35, 37

―――, *f. Native*, on lead-glance from *Illobor*, 166

―――, *g. Pitch-ore*; produced from hardened copper-mulm at *Saſka*, 40

―――, *h. Red-ore*, or *Tile-ore*, or red *copper-mulm*, produced from mouldering decaying jaſper, 39, and found at *Oravitza*, 31

―――, *i. White arſenical*, 31

*Copper*, (ſmelting) in the *Bannat* deſcribed, 57

―――――, Propoſals of Mr. *Delius* to improve the operation, and to ſoften the copper by additional ſulphur, or a ſulphureous regulus, 63

*Copper*, (veins) At *Dognazka*; *Simon Judas*, erroneouſly called a ſtock, 51

In metallic rock incumbent on gneiſs, 51

Conſiſts of copper pyrites, white limeſtone, ſpar, agathe, and yellow or black garnets, 53

―――――, *Mary Victory*, in metallic rock, conſiſts of diſſolved mica and copper pyrites, 54

―――――, At *Golniz*, in horn-ſlate, running from weſt to eaſt, conſiſt of grey quartz, ſome ſpar, copper pyrites, and grey copper-ore, 176

*Copper* (veins) At *Moldova*, run in every direction, between grey argillaceous ſlate and a hanging of limeſtone, both incumbent on gneiſs, 44

―――――, At *Muckenthurmel* in gneiſs, 315

―――――, At *Newſol,* in argillaceous ſlate, running from north to ſouth, dipping from eaſt to weſt, between forty and fifty degrees, cut off by an oblique croſs joint of red irony argillaceous ſlate; conſiſt of common ſhiſtous clay, mixed with mica, quartz, gypſum and copper ores, which are auriferous in ſome places, 195, 196

―――――, At *Oraviza*, in the mountains of *Coſhowiz*, run between argillaceous ſlate and limeſtone, conſiſt of calcareous and ſelenitic ſtones, 28, 29

―――――, in the mountains of *Cornudilſa*, run in limeſtone, conſiſt of gypſum and phoſphoreſcent ſpar

―――――, At *Saſka*, run between marle mixed with baſalt grains on the hading, and limeſtone on the hanging ſide, conſiſt of calcareous or ſelenitic ſpar with ſome quartz, 33—35

―――――, At *Smolniz*, in blue

# INDEX.

blue glimmery argillaceous slate, parallel to each other, run in hour six, dip in seventy-five degrees, improved and altered in their direction by small crossjoints, consist of dark grey clay with quartz and copper-pyrites, 171, 172
————, At *Swadler*, in glimmery argillaceous slate, 176
*Corsars*, strolling merchants allowed to purchase gold from private mines and wash-works in *Transsylvania*, 126
*Crystallisations*, with inclosed water, or *crystalli enhydri* at *Kapnik*, 155
————, (quartz) great variety in the large veins at *Shemniz*, 189
*Csertes*; gold mines in metallic rock, 123
————, auriferous glass-ore in hornstone, 122
*Debrezin*; fossil and native alcaline salt, 4
*Delius (Cristoph. Trawgott)* proposals to soften copper by sulphur, or an additional sulphureous regulus, 63
*Dembsher (Francis)* examination of the gold-dust and and gold-washings in the Bannat, 83—93
*Deva*; its copper stock-work, 94
*Diamonds*, white and red, white and yellow in the imperial cabinet at *Vienna*, 226
*Dognazka*; veins, rocks and ores described, 47—56
*Ducca*; a generous character of a man, 10
*Electrical* evaporation of a pyritical vein, during a thunderstorm, 60
*Entomolithus paradoxus trilobus*; an undescribed species in the sandstone beds on the *Harz-forest*, 251
*Facebay*, near *Zalathna*. The *Maria Loretto* gold-mine consists of a small stratified auriferous sand-stock and gold-veins, 110—117
*Feld-spath* (red) in the gold-vein at *Nagyag*, 101
————, matrix of auriferous fallow silver-ore at *Kapnik*, 153—155
————, the redder the more auriferous, 155
————, with silver, common in the richer veins, 220—319—320
————, its red colour owing to iron, 155
————, in metallic rock, 127
*Felso-banya*; a mining place, mountains, veins, works and ores, 158—166
*Firing* of mines, at *Goslar*, *Schlackenwalde*, and *Felso-banya*, 161
*Fissures or Klufte*, smaller veins, less constant and wide in their run and dipping, 28
*Fluor* cubic; with inclosed native

# INDEX.

native sulphur, 160
*Fluor*, in rhomboidal prisms 160
———, blue ⎱
———, green ⎰ with
———, yellow quartz in a tin-vein in granite at *Zinnwald*, 309
*Freygold* (*Martin*) at *Freyberg*, said to have invented the blasting of mines by gun powder, 192
*Fridwalsky* (*P.*) at *Clausenberg*, author of a Nat. History of *Transylvania*, 106, 107, 146, 147
*Fuczes*; mines, rocks, veins, and ores, 124, 125
*Gang.* See vein
*Gang mountains*; German name of ancient metallic mountains, which by some philosophers too inconsiderately have been called primitive mountains, 302
*Garnets*, found with gold-dust in the Bannat, 71—91
———, in grey micaceous and argillaceous slate at *Bleystadt*, 267
———, (yellow and black) with eighteen or thirty-six points in *Simon Judas* vein at *Dognazka*, 53—56 in *Paul*'s lead mine in the hading of the former, 54
At *Dognazka* erroneously called yellow blende, 56
*Gelft*; common name of pyrites in some *Hungarian* mines, 217
*Gipfies*, in the Bannat of *Temeswar* and in *Transsylvania* commonly employed in gold washing, 76
*Glass ore.* See copper-ore; or native silver from *Joachimsthal* 272
*Gneifs*, an argillaceous rock or mixture of quartz, mica and white or reddish clay, 243
———, the mountains at *Catharinaberg* and *Graupen* in *Bohemia* consist of it, 228—312
———, the veins at *Catharinaberg* filled with it, 245
———, at *Sofka* cap'd with argillaceous slate and limestone, 42—52
———, under columnar basaltes, 228
———, incumbent on granites, 236, 205
———, insensibly connects and degenerates into micaceous and argillaceous slate; accordingly to be consider'd in respect to situation as a variety of argillaceous slate; but in respect to mixture it is a variety of granites, 229—236—247
———, degenerates at *Commotau* into common argillaceous slate, 247
———, white, silver-coloured, blueish and dark-coloured at *Presniz*, 248

*Gneifs*

# INDEX.

*Gneiss*, Silver-vein in it, 248, 312
———, is a common matrix of silver-veins, 312
———, contains, besides the silver-veins, tin-veins at *Graupen*, 312
———, copper veins at *Saska*, 42, 52 and at *Muckenthurmel*, 315

*Gold-dust and washings*; in the Bannat of *Temeswar* found in the river sand and beds, which are parallel to the turf, and consist of loam, rockstones, mica, garnets and fine irony-sand, incumbent on slate or brown sandstone and coals, 77, 78—84—87—91
———, in the Bannat not to be described to any gold-veins, since never-found sticking to any matrix, 86
———, its origin investigated, 92
———, its washings in the Bannat described and examined, 76
———, in *Transsylvania* found under the turf in a sandy stratum, incumbent on argillaceous slate, 136
———, all the brooks in *Transsylvania* carry gold-dust, 137
———, is washed by the *Wallachians* in *Transsylvania*, and produces every year about a weight of thousand pounds, 110

*Gold-ore*, according to a popular opinion of the *Transsylvanians*, found only immediately under the surface of the horizon, 121—130

*Gold-ore*. See *gold-dust*, *zinnopel*, and *veins*,

A. *Native*.
 1. In calcareous spar from *Staniza*, 129
 2. In radiate antimony, from *Staniza*, 129
 3. In grey scaly cobalt from *Staniza*, 129
 4. On blende and lead-glance, 127
 5. In black lead, 226
 6. On selenite from *Fuezes*, 125
 7. On auriferous pyrites and black hornstone, 129
 8. In grapes and pretended vegetable gold are gross impositions or mistakes, 225
 9. On zinnopel, 166

B. *Mineralized*.
 1. Lamellous, splendent, black-grey, or woven in feldspath, from *Nagyag*, 98—101 found with native silver, 101, 102, and orpiment, 102
 2. like scaly antimony from *Nagyag*, 102
 3. In pyrites, or sprinkled upon as *Spanish* snuff, from *Facebay*, 115

*Gold-ore*,

# INDEX.

*Gold-ore,*
C. *Auriferous ores or Substances*
 1. Calcareous auriferous earth, found by nodules in the veins at *Kapnik,* 154
 2. Red silver-ore auriferous, from *Trsztyan,* 129
 3. Glass-ore auriferous in horn-stone, 122
 4. Lead-glance auriferous, in white clay, at *Kisbanya,* 129
 5. Pyrites auriferous, in blue clay, from *Herzigan,* 128
 6. ——— in black hornstone, from *Ginel,* 129
 7. ——— in black clay, from *Cajonel,* 129
 8. ——— on quartz, from *Cajonel,* 129
 9. ——— on blende from *Cajonel,* ibid

*Gold-veins,* running
A. *In metallic rock.* See *metallic-rock*
 1. At *Abrud-banya* near *Zalathna,* 117 In the *Kirnik-mountain* they are thin and short; being firstly vertical and deaf for eight fathoms; then dipping till they become soaring and auriferous for about two fathoms, when they turn again and break of, 118
 2. At *Boicza* the metallic rock is covered with limestone, and the veins consist of blendish and auriferous lead-glance 127

*Gold Veins in metallic rock.*
 3. At *Csertes,* the metallic rock cap'd with slate, 123
 4. At *Kapnik,* run from north to south, dip from west to east; consist of red feld spath, fallow auriferous silver-ore, auriferous quartz and clay, 152—154 Their auriferous quality diminishes in the depth, 155, increases wherever antimony appears, 154
 5. At *Kremniz,* run from south to north, consist of solid quartz, auriferous red and white silver ore and auriferous pyrites, 194
 6. At *Nagyag,* the metallic rock covered with red clay, run from north to south, consisting of red feldspath and white quartz 97—101
 7. At *Rota,* run between greenish calcareous and white metallic rock; contain blende, lead-glance and native gold, 155, 156
 8. At *Sargo-Banya,* consist of auriferous silver and lead, 165, 166
 9. At *Shemniz,* run from north to south, consist of quartz, lead-glance, and zinnopel; produce gold,

# INDEX.

gold, filver, lead, 181 —190

*Gold-veins, in metallick rock.*

10. At *Topliza,* in metallic rock cap'd with flate confift of auriferous quarts native gold, auriferous filver, and lead-ores; run from fouth to north, immediately under the turf, 123, 124

11. At *Ui-banya,* or *Konigſberg,* run between metallic-rock and granite; confift of grey quartz and auriferous pyrites, 200

B. *In Hornſtone.*

1. *At Facebay.*

a. *Sigismund gallery,* confifts of quartz, hornftone, auriferous pyrites, auriferous clay, and gold mineralized with pyrites, 110—112

b. *Maria Loretto,* confifts of two parallel veins and a ftratified auriferous fandſtone ftock, containing mineralized gold in pyrites, 110—113 —117

2. At *Felſo-banya,* run in grey hornftone and metallic rock, which is under the former; confift of zinnopel, which contains gold, filver and other-metals, 159—164

*Granite,* under columnar bafaltes, between *Lowoſiz* and *Topliz,* 228—under gneifs 225—236 Is the undermoſt ſtratum of the higheſt mountains and deepeſt mines in *Hungary* and *Tranſſylvania,* 202,—203

*Granite,* appears no where incumbent on or alternating with other rocks, 204

however may be incumbent in unexplored depths on fimpler rocks, hitherto undifcovered, 204

does not contain any metallic veins in *Hungary,* 204

contains many tin-veins in *Bohemia,* which confift of granite at *Platte,* 263 at *Catharinaberg,* 245—at *Zinnwald* 307—309

The tin-ſtock at *Schlackenwald* confifts of granite, ſtriped with pure quartz and tin-ore, 291. This tin-ſtock furrounded with gneifs ibid

A piece of granite with a fragment of flate ſticking in it, 207, 208

Was in a ſtate of paſte and ductility when in fome places flate was accumulated on it, 207, 208

A grey and reddiſh fpecies near *Kladraw* breaks and naturally fplits into cubical and rhomboidal forms and priſms,

# INDEX.

prisms, 297
Granite, may be produced by modern revolutions, as appears by the lava's which resemble to granitello, 289
　　If its feld-spath particles moulder into clay, it is called gneiss, 231, and petuntse, 232
　　Produces by mouldering argillaceous slate and gneiss, 229—236 —247. Clay more or less mixed with quartz, mica and feld-spath particles, 229—236—247. Perhaps china-clay, 230—235. quartzous sand, 231
　　Contains iron-veins at *Platte*, 262
　　With black crystallized sherl, 296
Granitello, resembling to some lavas, 289
Gyalter, an iron mine, 131 consists of small stocks or nodules in grey and brown argillaceous slate 131
Gypsum, constantly found in and about the rock salt-mines, 144
―――, pellucid striped white in the rock-salt at *marmaros*, 165
Habichwald near *Cassel*, coal-mine described, 305
Halotrichum Scopoli, seems to be an efflorescence of vitriol, 223
Herrings, fished now and then in fresh water in the Szamos-river in *Hungary*, 166
Horn-slate, (*Corneus Wallerii*) consists of quartz closely mixed with mica and clay, incumbent on granite, contains in *Hungary* some thin metallic veins, 205
　　Never contains any quartz-veins, and is found near *Wilkishen* in *Bohemia* in horrizontal beds, 296, 297—moulders into clay, 130—240 is a variety of grey micaceous slate, and called in Bohemia pochwacke, 260
A vein of blackish slate running in it at *Platte*, 261
Quartz and copper-veins running in it at *Golniz*, 176
Antimonial-veins running in it at *Rosenaw*, 177
Horn-stone (*Corneus Wallerii*) at *Facebay*, incumbent on argillaceous strata, contains rich gold-veins, and seems to be produced by modern floods or revolutions, 212, 110, 114 in *Hungary* never found incumbent on lime, 213 contains auriferous silver-veins at *Csertes*, 122
And problematical round holes, 114
By P. Fridwalzky erroneously called calcedony, 119
――― a. Black, with native gold, 129
――― b. Grey (petrosilex incumbent on metallic rock,

# INDEX.

rock, with auriferous silver and zinnopel-veins, at *Felſö-banya*, 159---164
Grey, and flint-like, fills a vein at *Joachimſthal*, 260
——— c. Red, ſemipellucid, flintlike, in the northern veins at *Joachimſthal*; the matrix of the richeſt ſilver-ores, 260—272
——— d. *White*, ſchiſtous and rocky, reſembling to calcedony, and flintlike, ſtratified, with petrifactions of corals; near *Lebotka*, 194
*Hot-wells*, at *Ofen*, 4
———, at *Shemniz*, in limeſtone, produce calcareous tophus with iron ocher, 193, 194
*Hyſterolithus, alatus planus latior*, in the ſand-ſtone beds on the *Harzforeſt*, 251
*Jacquin*. At *Vienna*, 226
*Jaſper*. See zinnopel,
———, *Red* or deaf zinnopel, found in micaceous clay-ſlate near *Shemniz*, 185
———, *Brown*, an undeſcribed ſpecies moulders into red copper-ocher, 39
*Joachimſthal*. The mountains black ſlate, 254
———. The veins unaffected by the direction of the valleys, 257
———. The mines extremely deep, 256
*Joints* or croſs-fiſſures, called *Kleins* in *Hungary*, 171
*Iron-ore*, incruſtated with blueiſh dripped calcedony from *Poinik*, 199
*Iron-ore*, ſand with gold-duſt in the Bannat and in *Heſſe*, 77—91
*Iron-veins*, at *Orpes*, ſoaring between limeſtone and incumbent white argillaceous ſtone, 250
———, in ſlate, at *Stooſs, Krumbach, Abavira, Rhoniz, Poinik*, 175--177 199
———, in granite, or between granite and ſlate, at *Platte*, 262
———, uniting with arſenical ones, ſuch as tin-veins, produce ſilver in the *Saxonian* mines, 264, 265
——— ore appears in the upper-drifts of the tin-vein at *Gotte-gab*, 265
*Kaolin*, is the ſubſtantial earth of granites, 232
*Kapnik*-mountain-rock-veins, and rich gold-ores, deſcribed, 152--154
*Kirnik-mountain*, contains a great number of ſhort and rich gold-veins, 117
*Klein*, a popular *German* name uſed in *Hungary* for the croſs-joints of the main vein, 171
*Kluft*. See *Fiſſures*.
*Kneiſs*. See *Gneiſs*.
*Koczian*'s obſervations on the gold-waſhings in the Bannat, 76
*Koleſeri (Sam.) Auraria Romano Dacica*; or account of the ancient *Roman* antiquities and mines in *Tranſ-*

# INDEX.

*Transsylvania,* 106

*Kremniz,* a great mining place, mountains, veins, ores and works described, 194

*Lava's,* some resembling granitello, 289

———, *vitreous* or pretended *Iceland*-agathe;
*a.* black.
*b.* blueish, semi-transparent, in detached pieces near *Tockay*; called lynx or lux-sapphires in *Hungary,* 167

———, See *metallic rock, Nagyag,* and *volcanic* productions.

*Lead glance,* auriferous, 129—124

———, and with native gold in white quartz and blende at *Rota,* 155, 156

*Lead-vein*
A. In limestone; at *Moditska,* 199
B. In slate, 132—267
C. In metallic rock, at *Dognazka,* 47, 48
At *Topliza* and *Fuezes,* contains native gold, 124

*Limestone,* incumbent on clay, metallic rock and granite, with sea-shells at *Bogshan,* 62

———, incumbent on argillaceous slate, 24--28

———, constantly incumbent on clay, contains in *Hungary* some lead and copper veins, 206

——— often immediately incumbent on granite, 207

*Limestone,* a cinnabar-vein in it, 120, 121

———, (*scaly*) occurs in the *Simon Judas* copper vein at *Dognazka,* 53

——— seems to have had a different origin, and to be either ancient or accidental, 210, 290

———, the granulated and scaly destitute of petrifactions, 210

———, the accidental and modern limestone beds, containing petrifactions, are in *Hungary* destitute of metallic veins, 211

———, tophus, like stalactites in globular and columnar forms, produced near *Liptaw* by the waters coming from the higher *Carpathian* mounta is 198

———, copper-veins in it at *Oraviza,* 28, 29

*Lux* and *Lynx-Sapphire,* in *Hungary,* a popular German name of vitreous blackish and blueish semipellucid lava; erroneously called *Iceland-agathe,* 167

*Manganese,* red, crystallised in a hanging fissure of an auriferous zinnopel-vein, 161

*Marmaros-stones,* octangular alum-like quartz-crystallisations, 165

*Mercury-mines.* See *Cinnabar* and *Quicksilver-veins*

*Metallic-rock,* an argillaceous rock, mixed with mica, quartz,

# INDEX.

quartz, feldspath and basalt-grains, 33, 34
*Metallick-rock*, is so called in *Hungary* (or rather by Baron *Born*) because the richer mines of that country are constantly found in it, 33, 34
———, immediately incumbent on granite, 205 206
———, under the argillaceous slate, 181--123 --97--135
———, (grey) with sherl, quartz and spar-grains; the common rock at *Kremniz* and *Shemniz*, 181--191--123
——— contains gold and lead-veins, 123--125 --165--166
———, a variety, instead of mica mixed with litho-marga, as a wedge or stock in the common metallic rock, 189
——— (white) with litho-marga, contains veins of quartz and clay with auriferous silver, gold and antimony, 152--154--191
———, a variety, naturally split and broken, in flat regular pieces, near *Nagyag*; resembling to some lava's from the *Euganean-mountains*; seems to be a volcanic production, 133
———, a variety formed like bullets, found in the sound metallic rock near the *Theresia-vein* at *Shemniz*, 188

*Metallic - rock*, erroneously called sandstone at *Dognazka*, 54
———, contains gold-veins at *Abrud banya*, 117
at *Csertes*, 123
at *Kremniz*, 194
at *Nagyag*, 67-105
at *Nagy-banya*, 150, 151; likewise silver at *Sargo-banya*, 165, 166; likewise lead at *Rota*, 155, 156
———, contains copper-veins at *Dognazka*, 51
*Mica*, produced from clay, into which it moulders again, 229
———, white and dissolved in the copper veins and rocks at *Dognazka*, 54
———, (yellow) on cat-gold skirts, and incloses the tin-veins in granite at zinn-wald, 309
*Mispickkel*, uniting with a silver-vein improves it, 318
*Moldova*; veins, rocks and ores described, 44
*Mountains*. Their distinction into *primitive* or *primogenial* and *secundary* or *accidental* modern mountains, is merely relating to the different times and accidents of their origin, and implies no real difference in their substances, 302
———, *incumbent* or *accumulated*, are modern in respect to the lower strata on which they are accumulated, 310
*Nagyag*, properly called *Sekeremb*,

# INDEX.

*keremb*, a gold-mine and mining place, 96—105

*Nagy-banya*, mining, place, auriferous silver in metallic rock, 150, 151

*Newsol*, mining place, mountains, rocks, veins, ores and works, 195—98

*Nummularii*, or *lapides numismales Transsylvaniæ*, from *Torda*, 143

*Ofen* petrifactions and hot-wells, 3, 4

*Orpiment*, or crystallised arsenic, in a vein of blue clay in slate, incumbent on metallic rock, 195

———, with mineralized gold, 102

———, with native sulphur, 160

*Pest*, a city in *Hungary*, 3

*Petrifactions* of sea shells in limestone near *Ofen*, —near *Clausenburg*, 146—149

———, in white stratified hornstone or calcedony, at *Lehotka*, 194

———, in a trap-vein, running in old slate-mountains, a petrified tree, called the *diluvian-tree*, 265

———, in a zinnopel vein at *Shemniz*, madrepores or porpites, 184, 185

*Petunse*, is decaying granite, in which the feldspath moulders into clay, 232

*Plajashes*, a national militia in the Bannat of *Temeswar*, 10

*Pocket-work*, used before the blasting of the rocks, 188

*Porphyry* (red) (veins) uniting with the veins at *Aberdam*, improve them with silver, 260

———, or large combs at *Joachimsthal*, quicken the vein, 261, 263

*Primitive* or *primogenial mountains*, a precarious denomination, implying only that they are anterior in time to the origin of the incumbent more modern strata and mountains, 302

———, caverns discovered in them at *Joachimsthal*, 267

———, petrifactions, but very scarce, in the veins which run in them, 265 —184—185

*Pyrites* in a tin-vein at *Zinwald* 309

——— quickens the tin-veins at *Gottesgab*, 266

———, auriferous, in clay, hornstone, quartz and blende, 115—129

———, called *gelft* by the German miners in *Hungary*, 217

*Pyritical vein* gives an electrical flame during a thunderstorm, 60

*Quartz* quickens the copper-veins at *Smolniz*, 172

——— (*auriferous*) and lamellous, 191

———, with native gold, 129, 130

———, (*crystallisations*) pointed on both ends, 160

———, octangular alum-like

# INDEX.

like from *Marmaros*, 165
———, in a tin-vein in granite at *Zinnwald*, 309
———, (*fat*) in the gold-vein at *Nagyag*, 101
———, (*grey*) vein, between granite and metallic rock, contains gold-pyrites at *Konigsberg*, 200
———, (*irony*) auriferous, 217
———, (*milky*) with native sulphur, 160
*Quickfilver-veins* at *Dumbrava*, near *Zalathna*, run in argillaceous flate and fand-ftone, confifts of quartz, fpar and found fcaly cinnabar, 120
At *Boboja*, run in limeftone, contain granulated cinnabar, 120, 121
*Raizes*, a *Sclavonian* tribe, calling themfelves *Srbi*, inhabiting *Servia* and the Bannat of *Temefwar*, 14
———, their language *Sclavonian*, 14
———, their character, 22
———, ufe the *Greek* alphabet, 23
*Rafpe's* (R. E.) defcription of the *Blockfberg*, 231
———, of granite-fands and clay, 231, 232
———, on the origin of china-clay, kaolin and petunfe, 232
———, on the origin of quartz fand, 232
———, defcription of marine fand-ftone-beds with unknown petrifactions, incumbent on the higher flate-mountains of the *Harz-forest*, 250
———defcription of the coal-beds on the fummit and around the *Habichwala* in *Heffe*, 305
*Rock-falt*, at *Marmaros* furrounded with micaceous clay-flate, 165
———, at *Torda*, incumbent on argillaceous flate, cap'd with limeftone beds, 140
———, confifts of ftratified falt, 141
———, gypfum and alabafter found between the falt-ftrata at *Marmaros*, are common near the falt-mines, 144
———, with included water-drops, 143
*Romun*, name of the *Wallachians* in the *Bannat*, *Tranffylvania* and *Wallachia*, 14
*Rofe-fpar*, fpecies of calcareous lamellous red-fpar, peculiar to a mine at *Joachimfthal*, 260
———, with red filver-ore, 272
*Saalband* the fimbriæ, out-fhirts or fide-covering of metallic veins; not conftantly appearing fmooth and polifhed as flickon-fides, but often grown to the fides of the mountain-rock, 155
———, of the tin vein in granite at *Zinnwald*, confift of yellow mica or cat-gold, 209
———, many veins, as the tin-

# INDEX.

tin-veins at *Graupen* in gneiss have no skirts, but are immediately grown to the sides, 313, 314

*Sal-ammoniac* manufactory at *Bronsvic*, 145

*Salt native*. See Rock-salt.

*Sandaraca*. See sulphur.

*Sand (quartz)* on the sea-shores and the plainer countries probably, produced from decayed granite rock, 232, 233

*Sandstone*, incumbent on argillaceous slate and metallic rock, 195

———, with red native sulphur, 195

———, stratified mineralized gold-pyrites in a stock at *Facebay*, 113

———, surrounds in *Hungary* the nobler metallic-mountains, 211

———, incumbent on limestone, 211

———, destitute of metallic-veins, 211

———, accidental modern slate-beds often incumbent on sandstone, 211

———, with petrifactions, 139

———————, of unknown sea-shells on the higher metallic and ancient mountains of the *Harz-forest*, 250

*Sapphire*, (Lynx or Lux) a vitreous semi-pellucid blackish and blueish lava from *Tockay*, 167

*Soska*. See veins.

*Saska*, copper-ores described, 37—41

*Sezugass* (Baron) 9, 10

*Selenite*, with native gold, 125

*Shemniz*; mining place, mountains, rocks, veins, ores and works, 180—194

*Sherl (crystallised)* in metallic rock, 123—181—191

a. *blue*, columnar, hexagonal or polyedrous, truncated, on copper-ore from *Saska*, 40

b. *black*, in *Bohemian* basaltes, 228
in granite, 296

c. *green*, in *trap* at *Joachimsthal*, 263, 264

*Skalka* (composed of the Hungarian article *is*, and the word *Kalika* a point or summit) the name of the top of a mountain in Hungary, which produces red native sulphur in sandstone, incumbent on clay-slate and metallic rock, 195

———, name of a sandstone quarry on the summit of the *Harz-forest*, 250

*Silver-ore*,

a. *Blachmann* signifies at *Kremniz* white silver-ore incrustating quartz, 219
At *Shemniz* it has another signification, 219

b. *Brush-ore*, a species of native silver from *Joachimsthal*, 271

c. *Fallow*, auriferous, in feldspath from *Kapnik*, 153

d. *Glass-*

# INDEX

*d. Glafs-ore*, in cubic forms, 218
*Silver-ore*, is mineralized with fulphur, 218—272
*e. Goofe dung-ore*, 219, 220
*f. Mulm*, 220
*g. Native* filver. See *Brufh-ore*.
 from *Joachimfthal*, 271
 with mineralized gold at *Nagyag*, 101, 102
 on pyrites, 218
*g. Native-filver* dendritical, from *Budweifs*, 316, 317
*h. Plumofe*
 1. *Grey*, in quartz, 219
 2. *White*, in irony quartz, 219
*i. Red.*
 1. *auriferous*, 124—129— at *Kremniz*, 218
 at *Topliza*, 124
 2. *Cryftallifed*, at *Joachimfthal* ruby-coloured and pellucid, 257
 from *Saxony*, fomewhat darker, 257
 from *Andreafberg*, darker, 257
 on rofe-fpar from *Joachimfthal*. 269—272
 on cobalt, arfenical pyrites and red horn-ftone from *Joachimfthal*, 272
 3. Dendritical and in globular lumps, 218, 219
*k. White*, auriferous at *Kremniz*; and called *Blachmann* when incruftating quartz, 219
*Silver-veins*. Iron fiffures in *Saxonia* uniting with arfenical-ones, fuch as tin-fiffures, produce filver, 264, 265—313
*Silver-veins*, by uniting arfenical ones improved, 318
A. In *gneifs*, at *Prefniz*, 241—at *Graupen*, 312
 Gneifs being a common matrix of filver, 512
B. In *Hornftone*, at *Cfertes*, 122
C. In *metallic-rock*. At *Kremniz* run from fouth to north, confift of folid quartz, and auriferous red and white filver-ore and pyrites, 194
 At *Nagy-banya*, with auriferous filver, 150, 151
 At *Shemniz*, run from north to fouth, confift of quartz, lead-glance and auriferous jafper or zinnopel, 181—190
D. In *black clay-flate*. At *Joachimfthal*, ftill quick in a depth of 350 fathoms, 254—257
 At *Claufthal* in the *Harz-foreft*.
 At *Weiperth* in *Bohemia* between gneifs and incumbent flate, 253
E. In *grey* and *blue clay-flate*, at *Ratieborziz*, 316
*Slate*, fragments of flate in granite, 207, 208
A. *Argillaceous* or *clay-flate*; of *the old-metallic* and pretended primitive mountains, in refpect to its mixture, the fame as that which is found in modern mountains; but is different in refpect of antiquity

# INDEX

*Slate*, A. *Argillaceous.*
quity and origin, . 289
Produced from granite and gneiss mouldering and decaying into clay; as appears from its being immediately connecting in the same mass with gneiss, 229—247—by petroleum changed into coals, 305
At *Kladraw* in *Bohemia* breaks and broken in regular cubic forms, or in rhomboidal prisms, 297
incumbent on granite, 47
incumbent on metallic rock, 195
under limestone, 24—28
contains iron veins and stocks, 192—177—131
lead-veins in it, 192—132
copper-veins in it, 206—at *Newsohl*, 195, 196
At *Oraviza* under a hanging of limestone, 28, 29
silver veins, 253, 254—257—316

B. *Blue*, micaceous *clay-slate*, contains the copper-veins at *Smolniz*, filled with dark grey clay and quartz, 170—172

C. *Grey*, micaceous *clay-slate*, is a variety of horn-slate, 260
forms and fills a vein in horn-slate, 261
incumbent on granite, contains at *Aberdam* silver and cobalt-veins, and sometimes tin-veins ascending from the deeper granite, 259
At *Dognazka*, contains copper-veins, 47

C. *Grey, micaceous slate.*
incumbent on gneiss, covered with limestone. 44
consists of grey clay and mica, 240—243
contains iron-veins at *Stoofs*; and between them some nests of copper-ore, 175
At *Moldova*, 44
copper-veins at *Swadler*, lead and quartz-veins at 176
*Bleystadt*, 267
D. *Horn-slate* See *Hornslate*.
E. *Modern argillaceous slate*, incumbent on sandstone and lime, caps the coal-beds in *Hungary*, 211
is vitriolic and furnishes the alum-works at *Commotaw*, 247
*Slickon-sides*, a species of smooth and polished skirts, fimbriæ and coverings or vertical joints in metallic veins, 155—See *Saalband*
*Smolniz*, mining place, 169—175
*Spar*, calcareous, with native gold, 129
———, (Rose) at *Joachimsthal*, with red silver ore, 260—272
*Srbi*, name of the *Raizes* in the *Bannat* of *Temeswar*, 14
*Stalactites arsenical*, or arsenic calx dripped in a stalactical form, from *Joachim-sthal*, 258
A problematical species, light, red and yellow as amber; vitreous and glossy;

# INDEX.

fy; refifts the acids; gives no fmell when burnt; found at *Felfo-banya* in the zinnopel gold-and filver-veins, 259

*Stalactites*, blueifh calcedony dripped as ftalactites, 199

*Stocks* or *Stockworks* at *Deva* and *Dognazka*, erroneoufly fo called, as confifting of many coinciding and uniting copper-veins, which run in metallic rock, and micaceous clay flate, having a determined direction and dipping, 50—53—94, 95

*Stock* at *Faccbay*, in grey hornftone, confifts of a ftratified auriferous cone or ftock, containing gold mineralized in pyrites, 113 at *Schlackenwald*, in gneifs, confift of granite and tin-ore; have the form of large inverted cones; are croffed by fome deaf and fome metallic veins, which feem to continue in the furrounding gneifs, uncertain whether having regular inclofures or flickonfides (*Stocfheider*) as the tin-ftock at *Geyer* in *Saxonia*; feem to be the fummits of old granite-peaks, furrounded with gneifs by modern revolutions, 269 --271--291

*Strata*, (*ancient*) granite, argillaceous rock, limeftone, 236—238

———, (*modern*) thinner and accidental, confift of clay, flate, marle, fand, lime, 237--247

———, (*marine*) on the fummit of fome of the higheft argillaceous and metallic mountains in the *Harzforeft*, confift of fandftone and petrifactions, 250

*Sulphur*, for its deftroying the iron, recommended in the refining of copper, 63

How feparated from the pyrites at *Swelniz*, 174

———, (red) cryftallifed or fandaraca in white cryftallifed quartz at *Felfobanya*, 160

———, on yellow orpiment, 160

———, in cubic fluor, 160

———, native in fandftone, 195

*Temefwar*, city, unhealthy, 11

*Tin-ore*, white, from *Schonfeld*, refembles to white or greenifh fat quartz, 294

*Tin-ftocks* at *Schlackenwald*, confift of conical granite-lumps, furrounded with gneifs, 269--271--291

*Tin-veins* at *Aberdam* in granite, and thence afcending in the incumbent flate, 259

At *Gottefgab*, carry iron in the upper drifts; produce tin a middle depth, and may yield filver in a greater depth, as arfenical or tin-veins croffed by iron-veins croffed are conftantly obferved

# INDEX.

in *Saxony* to produce silver in the crossing, 264, 265

*Tin-veins,* at *Graupen* in gneiss, 312

At *Platte*, run in granite, consist of granite and tin-ore, 262, 263

At *Schlackenwald*, the tin-stocks consist of granite, inclosed in gneiss, 269

At *Schonfield*, run in gneiss 293

at *Zinnwald* in granite, nite, consist of granite, 307--309

At *Aberdam*, the veins in granite produce tin; those which ascend into the incumbent grey micaceous slate produce silver and cobalt, 258

*Tockay*, vitreous black and blueish pellucid lava, called thereabouts *Lynx* or *Lux-Sapphire*, 167

*Toplitz*, hot wells and coal-works, 310, 311

*Topliza*. See gold-veins.

*Tophus*, seems to be produced by the sediments of hot wells, 194

*Torda* Salt mines, 140

*Transsylvania*. Antiquities described by *Kolesori*, 108

——— Natural History, written by *Koleseri* and P. *Fridwalzky*, 106

———, called *Ardellia* by the *Wallachians*, 14

*Trap*, argillaceous rock; *a. blue;* striking fire with steel; unmetallic or containing but small deaf veins; cap'd with micaceous clay-slate, 153--206

*b. black,*

*c. grey* and

*d. greenish;* hardened irony bole; contains spar and sherl-grains; occurs at *Joachimsthal* in large vertical veins, running in black slate, 263, 264 under ground extremely hard; moulders in open air into saponaceous bole, which dissolves in water, 264

contained at *Joachimsthal* in the cow-vein a petrified and pretended antediluvian tree, which is described, 265

*Trsztyan* near *Fuezes.* Gold mine described, 125

*Tshavoja.* Lead mines in blue micaceous clay-slate, 195

*Vansha (Peter)* a generous chief of robbers, saved the Emperor of *Germany* from being taken by the *Turks*, 10

*Vein* and *Gang* synonima, 47

———, *Fissures* are smaller veins, having a less constant and steady run and dipping, 28, called *Klufte* in *Germany*, 28

———, *Combs* or *Wacken* are large vertical veins of porphyry, trap and hornstone, 262

———, *Stocks* or *Stockworks* are large masses of ore, or metallic rock, as nodules included and surrounded with the mountain-rock; they

# INDEX.

they have no direction or run to any part of the compass, having commonly the form of a large cone. See *Stocks* and *Stockworks.*

*Veins,* the veins run commonly parallel to the valleys, except at *Joachimsthal,* where they are unaffected by the valleys, 257, 254

Their *direction* or *run,* is determined in *Germany* by the compass or a dial; accordingly those which run from east to west, or *vice versa,* between hour three and nine are eastern or *Morning-veins*; and those which run from south to north, or *vice versa,* between hour nine and three, are called *Northern* or *Midnight-veins,* 257

———, Their *dipping,* or *inclination towards the horizon,* is determined by a quadrant or sector; and they get accordingly different names, 245

—— *vertical* or *standing-veins* are those which dip from ninety to seventy-five degrees.

*Sliding* or *slipping Veins* (*Tonnlegigt*) those which dip from seventy-five to forty-five degrees.

*Flat veins,* those which dip between forty-five and fifteen degrees.

*Soaring veins,* which dip from fifteen degrees to o

*Veins,* petrifactions in veins, 184, 185--265

See *Cinnabar, Copper, Gold, Iron, Lead, Quicksilver, Silver, Tin-veins. Saalband.*

*Veins (Cross)* or *Cross-Joints,* are called *Kleins* in *Hungary,* 171

———, are such veins as run in the same mountain, but in a contrary direction or contrary dipping to the main vein, so that they cross it either in the run or dipping. In the last case they make what they call in *Germany* a *falling-cross.* They are commonly filled with substances, clay, stones and ores, which are different from those in the main vein, and affect it in a various manner.

———, at *Catharinaberg, Joachimsthal* and *Shemniz* they quicken and improve it, 245, 246--261--182, 183

———, at *Smolnix* the main vein deaf, unless quickened by crossing ones, 172

———, their various effect upon the main veins at *Smolnix,* 171

———, at *Felso-banya* strike the main vein dead, 160

See *Cinnabar, Copper, Gold, Iron, Lead, Quicksilver, Tin-veins.*

*Ventilator* and air-conductors in the gold-mines at *Nagyag,* 100

*Vitriol*

# INDEX.

*Vitriol*, how separated from pyrites, 175
*Volcanic productions*, scarce in *Hungary*, 213
See *Metallic-rock*; *Lava*. *Sapphire*.
*Wacken*; German popular name for large veins of hard rocks, such as porphyry and trap, 262
*Wallachia*; by the inhabitants called *Zara-more*, 14
*Wallachians*, call themselves *Romun*, and seem to be remainders of ancient *Roman* colonies, 14
⸻, their language corrupt *Latin*, 14
⸻, agrees with the common *Italian* in many qualities, 14
⸻, their manners in the *Bannat*, 15
⸻, their religion the non-united *Greek*, 17
⸻, make use of the *Greek* alphabet, 23
⸻, in *Transsylvania* more humanized and industrious, 94--137
*Wolfram*; the fore-runner of tin in the tin-veins at *Platte*, which run in and consist of granite, 263
*Wood*, petrified and irony in the clay beds near *Ofen*, 252
⸻, petrified found in a trap-vein, which runs in old slate-rock at *Joachimsthal*, 265
*Zalathna*, metropolis of the *Wallachians*; a great mining place in *Transsylvania*, 108--110--120
*Zara-more*, *Wallachian*, name of *Wallachia* 14
*Zinnopel*; red auriferous jasper, contains gold, silver, lead, zinc and pyrites, 216--188
⸻, some looser parts and samples look as red boles, 216
⸻, chief rock of the veins at *Shemniz*, which run in metallic rock, 182--189
⸻, petrified, madreporæ or porpites found in its vein at *Shemniz*, 184
⸻, at *Felso-banya* the zinzopel-veins run in metallic rock and hornstone, contain auriferous silver, 159--164
⸻, with native gold from *Olalapos*, 166

# FINIS.

BOOKS *printed for* G. KEARSLY, *at* No. 46, *in* FLEET-STREET.

I. THE Gentleman's Guide in his Tour through France. [With a correct Map of the Post-Roads, &c.] By an Officer. Containing an accurate Description of that Country. Including Paris, Versailles, Fontainbleau, Marli, St. Germains, St. Cloud, and every public Building and Place worthy a Traveller's Notice. Lists of Lodging-Houses, Ordinaries, Places of Amusement, with their Prices; Stage-Coaches and Water-Carriages to different Parts of the Kingdom, with their Fares; and every other Particular necessary for the Information of strangers.

II. The Tour of Holland, Dutch Brabant, the Austrian Netherlands, and Part of France; in which is included a Description of Paris and its Environs.

III. Useful Hints to those who make the Tour of France. In a Series of Letters written from that Kingdom. By Philip Thicknesse, Esq.

These Letters (none of which were ever published before) contain some Account of the interior Police in general, and of Paris in particular, with a considerable Number of entertaining Anecdotes, relative to the first personages on that Part of the Continent.

*⁎* These three Volumes, which may be had separate or together, Price 3s each, will enable Travellers to make the Tour of France and the Low Countries; as they contain every Thing worthy the Attention of the most minute Enquirer; and will prevent, if properly attended to, the scandalous impositions too often practised by the Publicans upon the continent.——The last Article only is written by Mr. Thicknesse.

www.ingramcontent.com/pod-product-compliance
Lightning Source LLC
Chambersburg PA
CBHW030429300426
44112CB00009B/926